Managing Tropical Animal Resources

Crocodiles as a Resource for the Tropics

Report of an Ad Hoc Panel of the
Advisory Committee on Technology Innovation
Board on Science and Technology for International Development
Office of International Affairs
National Research Council

In Cooperation with the Division of Wildlife, Department of Lands and Environment, Papua New Guinea

NATIONAL ACADEMY PRESS
Washington, D.C. 1983

The National Research Council was established by the National Academy of Sciences in 1916 to associate the broad community of science and technology with the Academy's purposes of furthering knowledge and of advising the federal government. The Council operates in accordance with general policies determined by the Academy under the authority of its congressional charter of 1863, which establishes the Academy as a private, nonprofit, self-governing membership corporation. The Council has become the principal operating agency of both the National Academy of Sciences and the National Academy of Engineering in the conduct of their services to the government, the public, and the scientific and engineering communities. It is administered jointly by both Academies and the Institute of Medicine. The National Academy of Engineering and the Institute of Medicine were established in 1964 and 1970, respectively, under the charter of the National Academy of Sciences.

The Board on Science and Technology for International Development (BOSTID) of the Office of International Affairs addresses a range of issues arising from the ways in which science and technology in developing countries can stimulate and complement the complex processes of social and economic development. It oversees a broad program of bilateral workshops with scientific organizations in developing countries and conducts special studies. BOSTID's Advisory Committee on Technology Innovation publishes topical reviews of technical processes and biological resources of potential importance to developing countries.

This report has been prepared by an ad hoc advisory panel of the Advisory Committee on Technology Innovation, Board on Science and Technology for International Development, Office of International Affairs, National Research Council. Program costs for the study were provided by the Office of Technical Resources, Bureau for Asia, Agency for International Development, under Grant No. ASB-0249-SS-00-1026-00 and the Office of the Science Advisor, Agency for International Development, under Grant No. DAN/5538-G-SS-1023-00.

Library of Congress Catalog Card Number 83-061910

Panel on Crocodile Farming

EDWARD S. AYENSU, Director, Office of Biological Conservation, Smithsonian Institution, Washington, D.C., USA, *Chairman*

HOWARD W. CAMPBELL (deceased) U.S. Fish and Wildlife Service, Gainesville, Florida, USA

ARCHIE F. CARR, Jr., Department of Zoology, University of Florida, Gainesville, Florida, USA

F. WAYNE KING, Director, Florida Museum, Gainesville, Florida, USA

FRANÇOIS MERGEN, Pinchot Professor of Forestry, Yale University, New Haven, Connecticut, USA

MICHAEL G. MORRIS, Furzebrook Research Station, Institute of Terrestrial Ecology, Wareham, Dorset, England

HUGH POPENOE, Director, International Programs in Agriculture, University of Florida, Gainesville, Florida, USA

ROBERT M. PYLE, Species Conservation Monitoring Unit, Cambridge, England

SHELDON R. SEVERINGHAUS, Representative, The Asia Foundation, Taipei, Taiwan

Contributors

Papua New Guinea

CAROL GABARA, Wildlife Division, Department of Oro Province, Popondetta

WASSAM GABARA, Wildlife Division, Department of Oro Province, Popondetta

KAROL KISOKAU, Director, Office of Environment and Conservation, Waigani

NAVU KWAPENA, Division of Wildlife, Department of Lands and Environment, Konedobu

MIRO LAUFA, Division of Wildlife, Department of Lands and Environment, Konedobu

GREG MITCHELL, Crocodile Farm, Lae

LEO NING, Division of Wildlife, Department of Lands and Environment, Wewak

Other Countries

MIGUEL ALVAREZ DEL TORO, Instituto de Historia Natural, Departamento de Zoologia, Chiapas, Mexico

ANGUS D'A. BELLAIRS, Department of Anatomy, St. Mary's Hospital Medical School, University of London, England

PETER BRAZAITIS, Department of Herpetology, New York Zoological Society, Bronx, New York, USA

ROBERT H. CHABRECK, Louisiana Cooperative Wildlife Research Unit, Louisiana State University, Baton Rouge, Louisiana, USA

MAX C. DOWNES, The Game Conservation Centre, Melbourne, Australia

KARLHEINZ FUCHS, Hoechst Aktiengesellschaft, Frankfurt, Federal Republic of Germany

CLAIRE HAGEN, Reptile Products Association, New York, USA

RENE E. HONEGGER, Zurich Zoo, Zurich, Switzerland

TED JOANEN, Rockefeller Wildlife Refuge, Chenier, Louisiana, USA

JOHN LEVER, Koorana Crocodile Farm, Rockhampton, Queensland, Australia

WILLIAM MAGNUSSON, Instituto Nacional de Pesquisas da Amazonia, Manaus, Amazonia, Brazil

HARRY MESSEL, School of Physics, University of Sydney, Sydney, New South Wales, Australia

FRED PARKER, Kirwan, Queensland, Australia

ANTHONY POOLEY, St. Lucia Estuary, Zululand, South Africa

CHARLES A. ROSS, National Museum of Natural History, Smithsonian Institution, Washington, D.C., USA

K. B. SALE, UNDP/FAO Crocodile Breeding and Management Project, Hyderabad, India

LUIS VARONA, La Habana, Cuba

CHAROON YOUNGPRAPAKORN, The Samutprakan Crocodile Farm, Samutprakan, Thailand

UTAI YOUNGPRAPAKORN, The Samutprakan Crocodile Farm, Samutprakan, Thailand

MYRNA WATANABE, New York Zoological Society, Bronx, New York, USA

ROMULUS WHITAKER, Guindy Deer Park, Madras, India

* * *

NOEL D. VIETMEYER, Professional Associate, Board on Science and Technology for International Development, *Crocodile Study Director*

National Research Council Staff
F. R. RUSKIN, *BOSTID Editor*
MARY JANE ENGQUIST, *Staff Associate*
CONSTANCE REGES, *Administrative Secretary*

This publication is dedicated to the memory of panel member Howard W. "Duke" Campbell who devoted most of his professional life to the conservation of crocodilians. Dr. Campbell was Chairman of the Crocodilian Specialist Group of the Survival Commission of the International Union for Conservation of Nature and Natural Resources at the time of his death in 1981.

Preface

The panel that produced this report met in Papua New Guinea in May 1981. Its purpose was to consider the principles of the Papua New Guinea crocodile farming program and their implications for economic development and for the management and survival of crocodilians elsewhere.

Crocodiles are an integral part of the tropical fauna; they are ecologically important, biologically interesting, and, potentially, a renewable natural resource of considerable economic value. The panel hopes that through this report the possibility of saving and managing this animal throughout the tropics can be better assessed.

Members of the panel consulted officials of the Ministry of Wildlife and Conservation in Port Moresby and visited crocodile farms in Moitaka, Popondetta, and Lae. The panel is grateful to Karol Kisokau, Navu Kwapena, and Miro Laufa of the Division of Wildlife for arranging its itinerary and visits in Papua New Guinea. It also wishes to thank Yano Belo, Minister of Environment, for hosting an evening social at the Moitaka crocodile farm; Greg Mitchell and his wife Judy, who entertained the panel at their home in Lae and conducted a tour of their company's crocodile farm; and Wassam and Carol Gabara who acted as guides and hosts in Popondetta.

The Advisory Committee on Technology Innovation (ACTI) of the Board on Science and Technology for International Development, National Research Council, is assessing scientific and technological advances that might prove especially applicable to problems of developing countries. This report is one of a series that explores promising areas of science previously unknown, neglected, or overlooked. Current titles in the ACTI series on *Managing Tropical Animal Resources* include:

- *Water Buffalo: New Prospects for an Underutilized Animal* (1981)
- *Little-Known Asian Animals with a Promising Economic Future* (1983)

- *Crocodiles as a Resource for the Tropics* (1983)
- *Butterfly Farming in Papua New Guinea* (1983)

These activities are supported largely by the U.S. Agency for International Development (AID). Program costs for this study were sponsored by AID's Bureau for Asia, and staff costs by AID's Office of the Science Advisor, which also made possible the free distribution of this report.

How to cite this report:
National Research Council. 1983. *Crocodiles as a Resource for the Tropics.* National Academy Press, Washington, D.C.

Contents

1 Introduction 1

2 Crocodile Farming in Papua New Guinea 9

3 Conclusions 16

4 Regulations, Safeguards, and Research Needs 21

Appendixes 26

A Crocodile Farming Around the World 26

B Practical Crocodile Farming 34

C Selected Readings 43

D Research Contacts 45

E Biographical Sketches of Panel Members 49

Advisory Committee on Technology Innovation 53

Board on Science and Technology for International Development (BOSTID) 54

BOSTID Publications 55

1

Introduction

Crocodiles, alligators, caimans, and gavials* have existed for some 200 million years—much longer than mammals—but they are now disappearing at alarming rates. Of the 21 or so species of crocodilians distributed in the warm waters of the world, at least 18 are threatened with extinction in most of the countries where they are found.

Although some species, such as the American alligator, appear to be out of danger because of strict conservation measures, many of the others survive mostly in national parks, protected preserves, or a few breeding stations. This is true for the slender-snouted crocodiles of Africa and Asia, the saltwater crocodile of Australia and Southeast Asia, the black caiman and Orinoco crocodile of South America, the Chinese alligator, the Siamese crocodile, and other species.

Habitat destruction is a major contributor to crocodilian decline; each year more breeding areas are disturbed as swamps and marshlands are drained, rivers dammed, estuaries reclaimed, and riverine forests denuded. However, illegal poaching by tribal people with their simple but effective traps, snares, and set hooks, as well as professional hunters operating with power boats, spotlights, and modern firearms are also decimating the animals over most of their ranges.

To a large extent these animals are being destroyed because of their market value. Crocodile is regarded as the costliest and most fashionable leather in western markets. Since World War II, demand for crocodile leather shoes, handbags, luggage, wallets, watchbands, and other expensive luxury articles has far exceeded supply. Even small items such as purses and handbags sell for many hundreds of dollars each. For instance, a ladies' purse or handbag made from crocodile skin can com-

*Present-day crocodilians are grouped into three families: crocodiles, alligators and caimans, and gavials (gharials). The animals differ from one another only in minor characters such as shape of snout, arrangement of scutes, and dental features. This report focuses mainly on crocodile species, but its conclusions are generally applicable to alligators, caimans, and gavials.

mand prices as high as $4,000. A pair of men's shoes may cost from $500 to $900, and a wallet from $150 to $250.

The crocodile trade peaked in the mid-1960s, when world markets absorbed more than 2 million crocodile skins each year. Today it is still large. In 1979, for instance, 1,000,000 caiman hides and 300,000 true-crocodile hides entered international commerce. In 1981 the United States itself imported 100,000 hides.

International markets for reptile hides and leathers are centered in France, Italy, the Federal Republic of Germany, and Japan. France, the single largest buyer of raw crocodilian hides, uses an estimated 300,000 to 400,000 skins a year. The major buyers of finished crocodile leather products are Hong Kong, Japan, Thailand, Singapore, and the United States.

Unrestricted hunting and poaching for hides are wiping out the large breeding animals. Excessive hunting has a devastating effect on crocodile populations because their age distribution is like a pyramid: a small number of breeding animals dominates a large number of juveniles and hatchlings, most of which never survive to maturity. Such societies, in which the size of future populations depends on only a few animals, are highly vulnerable to extinction; once some of the mature members are killed the population can crash. And it takes a long time for a crocodilian population to rebuild because for most large species the females do not begin breeding until they are at least 8 years old.

Rearing Crocodiles

Although there may seem to be no future for many crocodile populations, the situation is not hopeless. With intelligent intervention and under good conditions they can recover rapidly. Mature crocodiles have no enemies other than man, and, given some care and protection, a small number of breeders can produce a huge number of progeny each year. Mature females of the various crocodilian species usually lay between 30 and 70 eggs each year, and under normal conditions most of these eggs hatch successfully. The key to conserving the population is to protect the few mature animals and their habitats. Then, because of their fecundity, crocodilians can rapidly build up large numbers of young.

This has been exemplified by the American alligator (*Alligator mississippiensis*). Ten years ago its future seemed doubtful, but the legal protection of the populations has brought a remarkable recovery. Numbers are now so high that two states have lifted the ban on harvesting alligators, and between 10,000 and 20,000 American alligator hides now enter commerce each year.

The past few decades have seen several other examples of successful crocodile-rearing projects. Later chapters of this book highlight the national program in Papua New Guinea. In addition, three successful government-operated farms exist in India (where all the progeny is returned to the wild because current Indian law prohibits commercial crocodile farms). A remarkable farm with more than 3,000 breeding animals operates on the outskirts of Bangkok, Thailand. Australia has four crocodile farms, and a few African countries now have crocodile farms that are already beginning to supply hides internationally. For example, in 1982 Zambia had two such farms, Zimbabwe, five, South Africa, four (with five more planned or under construction), and Kenya, one. Appendix A lists these and other countries that are initiating farms for various crocodilian species.

The early technical success of these projects offers the expectation that with an appropriate framework of safeguards and research, crocodiles might become a thriving resource for tropical nations. If such experiences can be replicated, crocodilians and their habitats may come to be considered as resources to be managed and treasured. This will require considerable investment, strict legislation and law enforcement, and international cooperation and research, as well as careful monitoring of the traffic in farmed hides. But national crocodile industries are a possibility, and they could result in thriving natural populations that are free from the danger of extinction.

Such prospects may also provide economic incentives for preserving the often-fragile ecosystem in which wild crocodilians live. Crocodile farming could play a part by slowing the uncontrolled draining of swamps and other wetlands that cover large areas of the lowland humid tropics of Asia, Africa, and Latin America.

Crocodiles as a managed resource could economically benefit remote areas of the lowland tropics. Villagers there often have few alternative sources of income and possibilities for economic development are limited. Because the human population is relatively sparse, few opportunities exist for local trading; even where fish are abundant the problems of marketing are formidable. Indeed, in some areas crocodiles may constitute the only readily saleable resource.

Crocodiles as Farm Animals

Well-fed crocodiles grow quickly. Under ideal conditions they may reach lengths of 1 m or more in a year and 1.5 m in 2 years. They are normally harvested in the third year when they reach about 2 m in length. In this time their value may have risen from about $5 to as much as $200.

Crocodiles have acquired a reputation as voracious feeders; investigations reveal this to be false. The animals actually have modest food requirements. Many hatchling animals have a food conversion rate of about 50 percent; that is, the crocodile adds 1 kg of weight for every 2 kg of food it consumes. Cattle, sheep, and pigs would have to eat 3-5 times as much food to achieve the same weight increase. After 2 years the crocodile's growth rate begins to slow down. During the third year the conversion falls to about 25 or 30 percent, which is still a high figure, and makes crocodiles probably the most nutritionally efficient land animal for commercial husbandry. Only the growth of some fish is comparable.

The high food conversion efficiency is due to the fact that crocodiles have low metabolic rates and are normally extremely lethargic. They are active only in short bursts, spend hours immobile, and move only about one-third as much as mammals. Moreover, being reptiles, they spend almost no food energy maintaining body temperature. They bask in the sun to keep warm and seek shade or water to cool off. For these reasons crocodilians can thrive in marginal habitats unsuitable for mammals or birds.*

Crocodile farming is also space efficient. As long as they are sorted by size, hundreds of juveniles or dozens of larger animals can be penned together in a small area. Indeed crocodiles often choose to pile up on top of one another in stacks.

Little is known about disease in reptiles. However, as farm animals, crocodiles have a major advantage: they produce antibodies readily and have few problems with external infections. In the wild it is common to find crocodiles missing limbs or tips of tails, with eyes gouged out, or enormous scars on the body. But the wounds heal readily, with little sign of infection.† This minimizes the need for veterinary services, a distinct benefit in remote village farms. Nevertheless, internal bacterial diseases, such as salmonella, can get out of hand and destroy a program by reducing growth rates, lowering hide quality, or killing the animals outright.

Crocodile Hides

It is the belly skin that is the valuable part of a crocodile, and the worth of a hide is determined by the size of the belly skin, the smallness of its scales, and the hide's general condition. (Holes, cuts, scars, and rot drastically reduce its value.)

Although international markets utilize any crocodile skin from 0.3 m to 6 m long, the most sought-after hides are not the biggest but the

*Investigations on the Nile crocodile showed that a pelican takes 3 days to consume food equal to its own weight, whereas a crocodile takes 125-160 days (Cott, 1961).

†The American alligator is being used at the University of Florida College of Veterinary Medicine as a model for studies on antibody formation.

moderate-sized ones from animals about 1.5-2 m long. These hides are approximately 25-50 cm in belly width.* Large hides, for example those more than 3 or 4 m in length, are suitable only for luggage and briefcases because their scales are large. Smaller hides, on the other hand, are suitable both for items such as shoes, handbags, and wallets and for larger items.

Internationally, the most desirable hides come from the saltwater crocodile (*Crocodylus porosus*). It has proportionally the smallest belly scales of any crocodile, it lacks osteoderms,† and on the side of its body the scales are uniformly small. The next most valuable hides probably come from Morelet's crocodile (*Crocodylus moreletii*), the American alligator, the Siamese crocodile (*Crocodylus siamensis*), and the Nile crocodile (*Crocodylus niloticus*).

Papua New Guinea

The rest of this report highlights the program in Papua New Guinea, where during the last 10 years the government has made crocodile rearing an organized industry, much as poultry farming is elsewhere. This program, which is beginning to establish crocodiles as a significant natural asset, is designed both to protect the wild populations and to integrate traditional uses of these reptiles into a scientifically managed hide industry.

In Papua New Guinea crocodile farming‡ has become the cornerstone for the economic improvement of some of the world's poorest people. It offers a means for bringing the rural poor into the process of economic development, and it can be blended into a traditional village structure where land and resources may be communally owned.

The projects are small and many have had operational difficulties, but they suggest that conservation and economic development can be not only compatible, but also mutually reinforcing. The innovative idea is not that crocodiles can bring in money, but that sound conservation can be blended with marketing crocodile skins, meat, and by-products.

An important aspect of this approach to crocodile conservation is that it is based on protecting the existing landscape and resources. It provides a tool for conserving the species in their own wild habitats so that

*A crocodile's total length is approximately 4-4.5 times the width of its belly.

†Osteoderms are deposits of calcium carbonate under the skin. They are undesirable because they dissolve away during the tanning process, leaving a pitted surface.

‡In the official terminology of the Convention on International Trade in Endangered Species of Wild Fauna and Flora (CITES), the Papua New Guinea program is "ranching" rather than "farming" because the young livestock are mostly culled from wild populations and are not bred on the farm.

survival will not depend on a few captive specimens living under artificial conditions. It requires none of the bush clearing, fencing, forage planting or pesticide spraying that domestic animals often demand—important advantages in an economic development project in a fragile tropical swamp or rain forest ecosystem.

The Papua New Guinea approach, then, provides an economic incentive for wildlife protection. Everyone—from the villager to the minister of trade—has a stake in keeping the wild populations healthy. Out of self-interest, in addition to natural respect, large numbers of people become the guardians of the resource and the habitat needed to keep it surviving and productive.

The world's major conservation organizations have given Papua New Guinea's crocodile program their stamp of approval. In 1976 a team of scientists representing the International Union for the Conservation of Nature and Natural Resources, one of the most prestigious conservation organizations in the world, inspected the program. As a result, Papua New Guinea was given special dispensation, and its crocodile skins can be legally traded internationally. For example, because of its endangered status the saltwater crocodile is banned from trade by the Convention on International Trade in Endangered Species of Wild Fauna and Flora (CITES). An exemption, however, is granted to Papua New Guinea in recognition of the fact that its crocodiles now are sufficiently well managed to sustain a skin industry without seriously damaging the wild stock.

This program serves as a model for nations of the Americas, Asia, and Africa where crocodilian resources are still unmanaged or managed poorly. Crocodiles are being destroyed so fast that within about five years Papua New Guinea and other countries that have organized crocodile farming operations may be the only ones supplying significant numbers of skins to the international market.

Although the principles developed in Papua New Guinea deserve international attention, the recipe will not be a cure-all for problems of rural development or crocodile conservation. Instead, the Papua New Guinea experience suggests that local social, political, economic, and conservation goals can become the impetus for a successful blend of village improvement and wildlife protection.

Previous pages. Feeding time at the Lake Murray demonstration crocodile farm. (Steve Raymer, 1981 © National Geographic Society)

2

Crocodile Farming in Papua New Guinea

As recently as the 1950s, crocodiles were abundant in Papua New Guinea. Hunting was a major occupation and was unrestricted. Some Australians and Europeans made fortunes by shooting thousands of crocodiles a year to make shoes and handbags in Europe and North America.

Although it was obvious that wild populations could not sustain such wholesale slaughter, the destruction continued. By 1967 both the saltwater crocodile (*Crocodylus porosus*) and the freshwater crocodile (*Crocodylus novaeguineae*) were threatened with extinction. By 1968, despite increased hunting, the yield of skins had dropped in half; along the easily accessible river systems, crocodile populations had been wiped out. By 1969 the saltwater crocodile had disappeared from much of its range throughout the country, and wildlife officers estimated that without protection most specimens of breeding size would be eliminated within five years.

But how could crocodiles be protected? Papua New Guinea is divided by mountain ranges, ravines, torrential rivers, forests, seas, malarial swamps, and more than 700 languages. It would take hundreds of trained wildlife officers to enforce a ban on crocodile hunting, particularly in the face of opposition from tribesmen who have traditionally harvested crocodiles for food, decorative items, and implements.

The challenge was given to the officers of the Wildlife Division. Under the leadership of Max C. Downes, these officials concluded that the best way to protect cocodile populations was to halt the slaughter of the large breeding adults and build up a new hide industry based on the increased numbers of young that would result.

Few young crocodiles ever reach breeding age in the wild. The tiny, virtually defenseless hatchlings are easy prey for large fish, birds, or other crocodiles. Almost all of them are killed by predators or by other natural causes, such as floods..

What was needed, the wildlife officers concluded, were incentives to

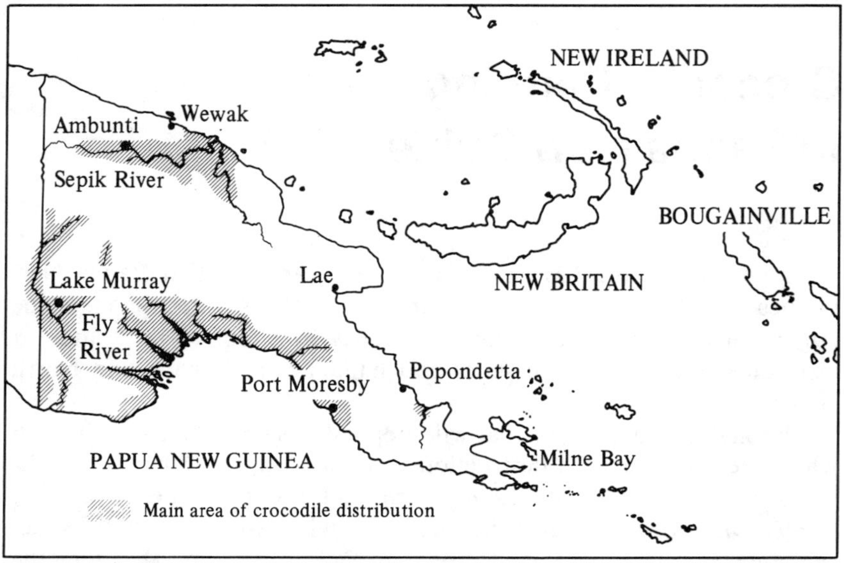

Main areas of crocodile distribution in Papua New Guinea.

make these smaller animals economically attractive, incentives to encourage local people to raise small crocodiles to commercial size. If that could be accomplished, hundreds of hatchlings that would normally perish could be utilized without endangering the wild populations' future. Villagers could benefit by selling skins while the vital breeding-sized animals were being left alone to provide more hatchlings. The system could benefit both the villagers and the vulnerable crocodile populations.

The idea was viable partly because many Papua New Guineans—particularly those of the Sepik and Fly Rivers—have ancient spiritual and cultural attachments to crocodiles. To them, the idea of handling and managing the animals is not unusual. Crocodile motifs are common in their art and they live in harmony with the big beasts and do not consider them dangerous pests to be eliminated.

Legislation passed by the Papua New Guinea government in 1969 capitalized on this tradition by making the villagers themselves the real force in crocodile protection. The law did not ban crocodile hunting, but instead banned the possession, sale, and export of skins larger than 20 inches (51 cm) wide. In this way, it protected breeding-sized animals while allowing for the harvest of juveniles. It also allowed a person to kill a crocodile if attacked (but barred the selling of the skin, if it were oversized).

In 1980, the legislation was supplemented by a law banning the export of small skins. Together, the bans on possessing large skins and exporting small skins have created a stimulus for gathering small crocodiles from the wild and rearing them to moderate size on farms. The legislation has been the impetus for crocodile farming.

Crocodile farming officially started in Papua New Guinea in 1972. In the late 1970s, it was extensively supported by a UNDP/FAO assistance program that provided personnel and funds for technical support and program management.Today there are about 300 small village farms* supplying a number of larger business groups that rear crocodiles.

Wildlife officers now teach crocodile farming, not crocodile conservation per se. They have introduced crocodile-rearing techniques to villagers all over Papua New Guinea. They help build pens and teach tribesmen how to care for the young reptiles, which are so vulnerable and timid that they can literally die of fright.

Government loans of up to US$10,000, along with matching development bank loans, are availble to help a farmer enter the crocodile farming business. The funds pay for pumps for changing the water in the pens and sometimes for an outboard motor used in gathering young crocodiles. Everything else a villager needs can be obtained from the forest, including materials for pen construction; a small farm can therefore be established inexpensively.

The Three Types of Farms

The government's crocodile management program recognizes three levels of operation: village farms (up to 300 crocodiles), small-business farms (up to 1,000 crocodiles), and large-business farms (more than 1,000 crocodiles).

A village crocodile farm consists of a small pen fenced with posts lashed together with vines. This stockade fence is about 1.5 m high and is sunk about 60 cm in the ground so the crocodiles cannot burrow out. Much of the enclosure is planted with grass, cassava, and banana trees to provide secluded areas where the animals, which regulate their body temperature by the warmth of the sun, can find shade. A shallow pool is excavated in the center.

*The numbers vary, since some villagers go in and out of production depending on their need for income, seasonal variation in river levels, the cost of fuel, and the availability of government extension agents.

Moitaka demonstration farm near Port Moresby. Meandering channels have proved more effective than large pools because they overcome the tendency of a dominant male to commandeer the water as his exclusive territory. (N.D. Vietmeyer)

These village farms are usually run by only one or two people. Many are little more than pens scattered in the remote bush for holding young crocodiles until a buyer from a larger farm comes around. Small crocodiles bring less money than medium-sized animals, but the villager avoids having to feed and care for them for a long period.*

The small-business farm usually consists of a group of enclosures (each about 6 m × 6 m) constructed of bush materials. It is typically located near an airstrip. It buys crocodiles from the village farmer and, in turn, supplies them to the larger farms, which sometimes dispatch aircraft to pick the animals up.†

*Because of operational difficulties, many village farms were abandoned in 1982. Lack of proper husbandy—despite government efforts—were the main reason for these difficulties. Most villagers now collect and hold young only until buyers from commercial farms arrive. However, they are still earning money from crocodiles, and the concept of fully functioning village farms remains valid for the future in Papua New Guinea, as well as for appropriate sites elsewhere.

†Special cardboard shipping containers have been devised. They can be folded to make cylinders of various diameters to fit crocodiles of different sizes.

Large-scale crocodile farms accommodate as many as 20,000 crocodiles and require a large investment. They serve to regulate the export of skins and are the major purchasers of live crocodiles from the smaller farms. During periods of drought, flooding, or diminished food supplies, the large-scale farms also act as emergency buyers. On the outskirts of Lae on Papua New Guinea's northern coast, there is a 100-hectare farm with nearly 8,000 crocodiles. It is associated with a poultry company, and the crocodiles are raised on the offal from the slaughterhouse.

Cutting up by-catch fish for crocodiles at the Moitaka farm. Crocodiles live mainly on fish, but insects, crustaceans, frogs, rodents, and slaughterhouse offal are also used. (N.D. Vietmeyer)

Village-level farm. Small crocodiles are timid, easily frightened, and often must be hand fed. (Division of Wildlife, Papua New Guinea)

Government Research and Extension

The Wildlife Division has constructed four demonstration farms across the country and one large research farm at Moitaka near Port Moresby, the capital city. After training at one of these, a tribesman can start his own farm alone or can call on the government for further assistance.

Moitaka is also the site of short courses in crocodile farming. Prospective farmers are brought in for several weeks' training. They learn how to build pens, to feed and care for crocodiles, to kill and skin them, and to prepare the hides for market. They also learn about the crocodile laws and the reason they were enacted. A farm at Lake Murray, in a remote and swampy area of the Western Province, serves the same purpose. It is built entirely from bush materials (see picture, pp. 6-7).

The Wildlife Division provides instruction books, profusely illustrated for the illiterate. The books include vivid descriptions of all phases of farming the animals.

Economic Gains

In 10 years, crocodile rearing has expanded remarkably in Papua New Guinea. It has already become the main source of income for the people of some swamp and river areas. The Ambunti area, for example, produces coffee and rice, but crocodile skins now bring in much of the area's income. By 1981 the farms nationwide contained a total of 30,000 crocodiles, ensuring a sustained production of at least 10,000 skins a year worth approximately US$1-2 million on the international market.

Because crocodiles are a familiar resource, villagers take to the program quickly. By contrast, introducing cattle or western-style crop raising requires massive education and training, in addition to some social and environmental disruption.

On government farms in Papua New Guinea, fish-fed crocodiles have increased their belly width by 25 cm per year and are ready for slaughter in 2-3 years when the width approaches 50 cm. The selling price of the skin is then between $100 and $200, depending on species, flaws, and size.

But skins are not the only product. A crocodile with a skin big enough to market can provide 20 kg of meat. The meat is white and is low in fat. Papua New Guinea is a net importer of meat, and crocodile farming is now augmenting local supplies. The large farm at Lae already sells frozen crocodile meat (including front and hind legs, tail steaks, ribs, and chops) both locally and on foreign markets. Some orders have come in from dealers in Paris who supply expensive French restaurants.

Crocodile farmers, Sepik River area.

3

Conclusions

Benefits of Crocodile Farming

Crocodile farming seems to be singularly appropriate for rural, isolated, lowland communities in the tropics. The land there is often unsuitable for conventional agriculture, and the people lead a tenuous existence or drift to the cities looking for work. In such areas, there are few opportunities for people to earn cash without drastic and expensive modifications to the environment.

Crocodile farming has many advantages over hunting the animal in the wild. For instance, crocodiles farms can:

- Permit government monitoring of the crocodile industry. (Hunters are more difficult to regulate since they work in remote areas, often undetected and crossing borders at will.)
- Yield a regular harvest of a specific number of animals of a selected size.
- Produce a standardized, premium product that better serves the needs of the international hide industry, making skins poached from the wild less desirable. (They may provide, for instance, a standard first-grade 1-1.5 m long hide rather than hides of mixed size and quality.)
- Reduce the wasteful losses of hides from improper handling, the fate of a high proportion of skins now brought in from the wild.*
- Educate the public about crocodile ecology and the animal's importance to the habitat and the local economy.

*In remote areas, salt may become scarce late in the season as it is used for hides. This often results in insufficient salt being used when the supply runs out and hides subsequently rot or "slip." In some isolated areas, as much as 25 percent of the hides are lost or downgraded because of improper curing for lack of salt. However, skins produced on farms, especially near urban areas or large villages, are usually properly salted; if salt becomes scarce, killing of the animals can be delayed.

Rather than breeding animals in captivity, Papua New Guinea's crocodile program is based on harvesting small animals from the wild and rearing them under controlled conditions. This makes crocodile farming more suited to village conditions and remote areas. It also creates incentives to maintain mature animals breeding in the wild and to protect their habitats. In this way, economic development is blended with conservation and with wetland habitat protection. (Division of Wildlife, Papua New Guinea)

• Provide sites for scientific studies on crocodilians. Studies conducted on alligators at the Rockefeller Refuge in Louisiana, USA, for example, have provided reproductive, nutritional, and growth data directed specifically towards developing efficient farming techniques.

For some farms the earnings from hides, meat, and by-products may be supplemented by tourism (through gate admissions and the sale of curios), as well as by selling eggs and young to other farms for breeding stock.

A long-term program of wise utilization of crocodiles can benefit governments by providing revenue from hides, curios, craftwork, and manufactured articles, as well as from export duties. Furthermore, in their natural state in parks and preserves, crocodiles are an important tourist attraction.

In an effort to preserve crocodile habitats, the Papua New Guinea program has encouraged the collection of eggs or young from the wild and has discouraged the breeding of crocodiles in captivity. This is because a reliance on the wild creates economic incentives to conserve crocodile habitats; if the habitats are drained for human settlement or conventional agriculture the farmers lose the source of their stock.

Contrary to popular impression, preliminary observations indicate that crocodiles benefit commercial fisheries. The animals are important links in the ecosystems of rivers and lakes and are often the largest inhabitants of the freshwater wetlands. Their movements inhibit the growth of aquatic plants in the waterways, and, in areas with prolonged dry seasons, some species maintain residual waterholes that benefit small aquatic organisms that would otherwise perish. In estuaries and lakes, crocodiles enrich the nutrient content of the water by converting terrestrial prey into feces that in turn feed invertebrates and fish.

Where crocodiles have been eliminated, reductions in the tonnage of fish caught for human consumption can usually be demonstrated. For example, in Brazil, Kenya, and India, a decline in the fishermen's catch has paralleled the decline in crocodiles.

Limitations of Crocodile Farming

Governments and individuals seeking rapid returns on investments should realize that a crocodile farming industry is not a get-rich-quick scheme. To build a stable national industry may require 10 years and an investment of at least $500,000 before it is biologically and economically successful.

Nevertheless, an organized industry is vital. A village crocodile-rearing pond is only profitable if there is someone to buy, grade, package, and ship the product with all its documentation. Services will be needed at all levels to advise on disease control, nutrition, skinning, and preserving the hides. In many parts of the world crocodile farms have been financial and conservation failures not just because of poor husbandry management, but also because of fiscal shortsightedness.

Selection of a suitable farm site is basic to the economics of the entire operation. Farms demand a steady supply of meat or fish to feed the crocodiles, and are most successful when located near a reliable source of inexpensive food. Some farms take advantage of offal from nearby chicken or cattle abattoirs; others use the fish by-catch from shrimping operations. In the absence of an inexpensive animal protein feed, the farm will have to raise its own food (tilapia is frequently used) or harvest it from the wild, both of which can be expensive.

Crocodile farms also require a steady year-round supply of clean water for the holding ponds and tanks. If this cannot be supplied by gravity flow from nearby sources, it must be pumped from wells or from nearby lakes or ponds. This, too, is likely to be expensive.

Despite the general hardiness of crocodiles, the farms must have access to veterinary care. Most disease problems stem from poor sanitation, low water temperatures, and poor diet, all of which can be easily corrected. But with large numbers of animals crowded together, disease problems, if not quickly diagnosed and treated, can wipe out the young captive animals in epidemic proportions.

Capturing and transporting large crocodilians is dangerous and difficult. Dealing with a large captive population of crocodiles of different age groups and sizes requires a great deal of experience.

Although crocodilians are common in zoos, successful breeding of these reptiles in captivity is so far a rare and remarkable event. However, researchers are now coming to understand the behavioral requirements for success. For instance, gravid females must have access to appropriate nesting sites, males must have ample space when they are penned in with other males, and juveniles and hatchlings must be separated from their parents and housed by size and feeding preferences. That prolific breeding can be achieved, however, is illustrated by the Samutprakan crocodile farm near Bangkok, Thailand, which reportedly has reared tens of thousands of its own animals and now aims for a population of 100,000 crocodiles by 1987.

Conservation

The worldwide shortage of crocodile leather is becoming more acute each year, and it will be many years before any output from farms can significantly reduce pressure on wild populations. Thus, farming should be only one aspect of an overall conservation program that includes total protection of some populations in national parks and sanctuaries. In addition, the conservation of natural wetlands is an important part of overall economic planning. If wetlands are lost, many wild species in addition to crocodiles will be affected.

In Australia, Asia, Africa, and the Americas, many crocodilian populations are poorly protected because governments lack the manpower or the will to enforce conservation laws rigorously, especially in the remote areas where the last remaining crocodiles reside. Because most wildlife departments in the tropics are short staffed and have vast areas to police, their efforts at wildlife protection are frequently ineffective. Moreover, some countries have been slow to introduce protective legislation for an animal that does not engender public sympathy.

Papua New Guinea's program offers one of the best hopes for saving all endangered crocodiles, alligators, caimans, and gavials. The methods developed there serve as a model for other nations. By providing an alternative, Papua New Guinea gives villagers the incentive to protect wild crocodiles that are breeding nearby so as to assure themselves of future supplies. The people themselves become the conservators of the local animals and habitats. In turn, watersheds, soils, and conventional agricultural development (including natural and forest products) can all benefit. The habitat is also preserved for many other wildlife species that share it, and genetic diversity can be maintained. Conversely, without a special incentive to conserve them, all these resources are normally degraded as a region develops.

4

Regulations, Safeguards, and Research Needs

The Papua New Guinea experience provides a model for other nations, but to implement such a program requires a foundation of legislation, government support, and legal safeguards. Prerequisites of any crocodile-farming program are an overhaul of legislation, strict law enforcement, and reciprocal laws with neighboring countries.

The enforcement of wildlife regulations is so inadequate in most countries that crocodile farming is open to abuse. Farms can front for illegal poaching operations, and hides taken from the wild can be intermingled with hides produced on the ranch or farm unless government enforcement is stringent and inspection frequent.

Moreover, the stimulation of world trade in crocodile hides through the sale of farmed hides might lead to increased poaching of wild crocodiles or eggs. Poachers have fewer operating expenses than farmers, and unscrupulous hunters and dealers can harvest hides, steal crocodile eggs or young, and subsequently sell them through countries that lack enforcement capabilities. This practice could be disastrous for countries where crocodile populations have almost disappeared.

Government Regulations*

Before any farming scheme is attempted, protective legislation should be in operation throughout the country. This should make it unlawful to kill, capture by any means whatsoever, disturb willfully, or pursue any crocodiles, or to collect or gather any crocodile eggs without a permit.

No crocodile eggs should be allowed to be imported or exported without a permit. No persons should possess, sell, buy, donate, receive consequent upon a donation, convey, keep in captivity, or display any live

*Suggestions of A. Pooley.

crocodiles without being the holder of a permit. And no person should be allowed to import or export any crocodile, dead or alive, or any portion of a crocodile, processed or not, from any country without a permit.

Before granting a license for a commerical farm, the government should investigate the applicant's land tenure and financial resources, particularly since the farm will have to operate for three to four years before producing crocodiles suitable for culling. The applicant's ability and experience in rearing crocodiles should be determined. A plan of the proposed farm, including details of water and food supply and the proposed methods of harvesting food, should be examined.

It is suggested that:

- No permit for egg harvesting should be issued until adequate rearing facilities have been prepared. The permit should state the name of the holder or his authorized representative, the annual total number of eggs allocated for harvesting, and the area where collection is permitted.*
- Permits should be issued on a year-by-year basis. The applicant should understand that the department may refuse to renew or issue further permits if the farm is not managed satisfactorily or if permit conditions have not been observed.
- The applicant should understand that the farm and all production records should be available for inspection by an official of the conservation department.
- The farmer should be required to submit periodic reports detailing the total number of nests raided and eggs harvested, the egg mortality, and the number of eggs hatched. Thereafter, the number of animals held in captivity, the rate of mortality and its causes, if known, and the number of animals sold or culled should be included in each report.

Furthermore, it is recommended that the permit holder release 5 percent of his annual crop of hatchlings in order to restock the natural habitat. In addition, a further 5 percent of the hatchling crop should be reared to a length of 1 m before being released, bringing the total release of young crocodilians to 10 percent of the annual crop of hatchlings.

The distribution of hatchlings and young reared animals should be supervised by the conservation department.

Unless government agencies monitor the wild populations of crocodilians being harvested for hides, eggs, or young, the farms themselves could become a major drain on those populations, leading to their extinction. Therefore, before any farming program is started, a survey of

*In some cases it is also important to specify a harvest time. Often it is best to take eggs laid early in the season because the female will then lay another clutch.

the breeding grounds should be undertaken to determine the number of nests available and those from which eggs can be taken with least danger to the wild population (for example, from nests on grounds likely to be flooded).

These breeding grounds should be fully protected; tourists on foot, in vehicles, or in launches should not be allowed to visit or disturb crocodiles during the breeding season.

International Safeguards and Cooperation*

The international traffic in millions of unmarked crocodilian hides and products poses one of the greatest obstacles to enforcement of national and international endangered species regulations. Hides and skins frequently cannot be traced to their source or country of origin. Legally harvested or farmed animals cannot readily be distinguished from those exported in secret from illegal sources.

The need for internationally acceptable methods of marking individual hides and products is critical. Traffic in illegal crocodilian hides and products will continue as long as law enforcement agencies lack the means to detect them easily.

In the United States a system has been developed in some states that enables conservation, police, and customs officials to monitor traffic in alligator hides. A conservation authority issues an official tag for each animal allowed by the license. All hides exported are tagged with a serially numbered plastic tag that cannot be removed without breaking it. The serial number is recorded on the export permit and with details of the buyer's and seller's name and address. This tag remains on the hide, right through the tanning process, until the hide reaches the manufacturer. Each tannery maintains a register of purchases that is available for inspection. This system also is being implemented in Zimbabwe and is worthy of trial in other countries.

The tagging of all hides and products for individual identification is an important safeguard. Other safeguards include:

- The use of engraved stamps or seals to authenticate legal licenses and export permits and make it more difficult for documents to be forged.
- Internationally accessible data and a retrieval system that allows law enforcement personnel to corroborate the authenticity of documentation and the origin of hides and products;

*Information in this section supplied by P. Brazaitis.

International Trade Restrictions

More than 100 nations have signed an international treaty restricting their importation of crocodile skins as part of a worldwide agreement designed to control international trade that could threaten the continued existence of species of wild plants and animals.

This treaty, known as the Convention on International Trade in Endangered Species of Wild Fauna and Flora (CITES), was established in 1973. Currently banned from importation (CITES, Appendix I) are the following endangered crocodilians:

Chinese Alligator	*Alligator sinensis*
Black Caimen	*Melanosuchus niger*
Rio Apaporis Caiman	*Caiman crocodilus apaporiensis*
Broad-Snouted Caiman	*Caiman latirostris*
False Gavial	*Tomistoma schlegelii*
Western African Dwarf Crocodile	*Osteolaemus tetraspis tetraspis*
Congo Dwarf Crocodile	*Osteolaemus tetraspis osborni*
African Slender-Snouted Crocodile	*Crocodylus cataphractus*
Siamese Crocodile	*Crocodylus siamensis*
Mugger Crocodile	*Crocodylus palustris palustris*
Ceylon Mugger Crocodile	*Crocodylus palustris kimbula*
Philippine Crocodile	*Crocodylus novaeguineae mindorensis*
Orinoco Crocodile	*Crocodylus intermedius*
Cuban Crocodile	*Crocodylus rhombifer*
Morelet's Crocodile	*Crocodylus moreletii*
Nile Crocodile	*Crocodylus niloticus*
Gavial	*Gavialis gangeticus*

Although signatory countries have agreed to prohibit trade in these species, authenticated captive-bred specimens can be exported and imported when appropriate scientific authorities in both exporting and importing countries can certify that the transaction will not be detrimental to the survival of the species.

A second category (CITES, Appendix II) limits (but does not ban) trade in crocodilians whose populations are threatened but not in imminent danger. These species are:

American Alligator	*Alligator mississippiensis*
South American Caiman	*Caiman crocodilus crocodilus*
Brown Caiman	*Caiman crocodilus fuscus*
Dwarf Caiman	*Paleosuchus palpebrosus*
Smooth-Fronted Caiman	*Paleosuchus trigonatus*
Johnson's Crocodile	*Crocodylus johnsoni*
New Guinea Crocodile	*Crocodylus novaeguineae novaeguineae*
Saltwater Crocodile	*Crocodylus porosus*
American Crocodile	*Crocodylus acutus*

Trade in the products of these species requires that exports be monitored by a scientific authority in the country of origin. If wild populations of a species appear to be dropping below the level that can maintain their role in the ecosystem, then exports are to be limited.

Again CITES will, in principle, give exemptions to countries that institute farming or conservation programs that in no way contribute to the further destruction of the populations. In 1979 Papua New Guinea, because of the success of its crocodile-farming program, was provided such an exemption for the saltwater crocodile.

Further information may be obtained from CITES, c/o IUCN Headquarters, Ave. du Mont Blanc, CH 1196 Gland, Switzerland.

- Monitoring agencies to record and publish market statistics, traffic, and trends;
- Laws limiting the sale of hides only to nations that cooperate in an internationally sanctioned program of safeguards; and
- Research funding to monitor populations and develop new marking and identification techniques. (For instance, the use of dyes, roll marking, and infusion of detectable chemical tracers has yet to be fully explored.)

Research Needs

There is urgent need for tannery owners, manufacturers, and conservation authorities to jointly work out the rational exploitation of crocodile populations. Commercial interests have reaped a rich reward over many years, and if the crocodile industry is to continue, its entrepreneurs must invest in management and conservation.

Clearly, research to improve farming techniques will be a wise investment for both commercial operators and the countries concerned. Surveys to determine population numbers and size as well as the structure of breeding stocks and recruitment rates are essential. Such surveys may indicate the need to establish sanctuaries to protect breeding stock and nesting grounds, or perhaps to ban hunting to allow populations to recover. A rearing program and restocking of suitable habitats might be necessary.

Appendix **A**

Crocodile Farming Around the World*

Experiences with crocodile farming in Papua New Guinea, the main subject of this report, are described in chapter 2. Here we summarize the status of similar efforts in other countries.

Australia

Four crocodile farms have been established in Australia, one in the Northern Territory and three in Queensland. To date, only the Edward River farm, operated by the government as an aboriginal development project, has developed a successful breeding program. There, seven-year-old saltwater crocodiles (*Crocodylus porosus*) hatched on the farm from wild eggs are now breeding and laying fertile eggs.

Asia

People's Republic of China

A farm for Chinese alligators (*Alligator sinensis*) has been established at Xuancheng, Anhui Province. Its purpose is to breed alligators for conservation, although the hide of this species is not in great demand because it has many osteoderms in the belly scales. Recently the government has expressed interest in establishing a farm for saltwater crocodiles in southern China.

Taiwan

Taiwan has one crocodile farm or rearing station, but it is too far north to breed its own stock, except in heated indoor enclosures.

*This chapter is based on material supplied by F. W. King.

Philippines

There are several buyers in the Philippines who maintain pens of crocodiles for short periods. None of these is a farm. A new experimental farm for the Philippine freshwater crocodile (*Crocodylus novaeguineae mindorensis*) was established by Silliman University in an attempt to preserve that endangered species and to promote an economic interest in crocodile conservation.

Micronesia

A farm for saltwater crocodiles has been established on Palau, where a small population of these crocodiles occurs in a brackish interior swamp. In the past, the government hired a hunter to reduce the population whenever the local people felt it had become sufficiently large to present a danger, about once a decade. Presumably the nuisance crocodiles will now end up in the farm. The farm, which has been in existence only for a year or two, earns money from tourist admissions as well as hide production.

Indonesia

A few crocodile-rearing stations have operated for several decades in Java, Sumatra, and Kalimantan. These have been stocked with eggs and young animals collected from the wild in Sumatra and Kalimantan. In the early 1970s, three such operations in Samarinda, East Kalimantan, closed down for lack of wild stock. At least one operation in Jakarta, Java, continues to survive, but with virtually no output of stock.

A survey of Irian Jaya (Indonesian New Guinea) in 1980 indicated a number of ranches in that region as well, but revealed that some were having difficulty obtaining stock because of overharvest.

Singapore

Singapore has a famous crocodile farm that figures prominently in tourism. It breeds some of its stock, but also obtains wild stock from all over Southeast Asia. Singapore has a thriving crocodile hide trade. Many buyers and several tanneries are located there.

Singapore is not a member of CITES and openly trades in any and all species of crocodilians.

Malaysia

There are several crocodile farms in West Malaysia (at Penang, for example) and at least one in East Malaysia (near Sandakan, Sabah). These started out as rearing stations relying on wild young, but have moved slowly toward breeding their own stock. The Penang farm depends on tourism to pay many of its expenses. The Sandakan farm is operated in conjunction with a duck and pig farm that supplies it with offal. Its stock consists of saltwater crocodiles. Until at least 1980, it had very little production from captive animals, but the owner is hoping to broaden his stock from them.

Sarawak (East Malaysia) also used to have several rearing farms. The present status of these operations is unknown.

Thailand

The Samutprakan Crocodile Farm was started in 1950 with 20 wild crocodiles and an investment of US$500. Today it is reported to be the world's largest crocodile farm, with about 30,000 individuals. About 3,700 of these animals, placed in eight separate breeding ponds, are used for breeding stock, and there are plans for a population of 100,000 by 1987. The Samutprakan farm opened to the public 12 years ago and now receives about one million visitors annually.

Most of the farm's crocodiles are from the two species native to Thailand, the saltwater crocodile and the Siamese freshwater crocodile (*Crocodylus siamensis*). It also has hybrids of the two, as well as the indigenous false gavial (*Tomistoma schlegelii*) and five exotic species: South American caiman (*Caiman crocodilus*), New Guinea freshwater crocodile (*Crocodylus novaeguineae*), Chinese alligator (*Alligator sinensis*), broad-snouted caiman (*Caiman latirostris*), and dwarf caiman (*Paleosuchus palpebrosus*). The farm has succeeded in breeding South American caiman (*Caiman crocodilus*); the other species are approaching maturity and it is hoped they will breed in the near future.

The farm sells crocodile meat locally, mostly to restaurants as a delicacy (for US$5 per kg).

The commercial and biological success of the farm is largely due to favorable conditions at Samutprakan. The temperature and humidity are high year-round, and low costs of labor and building materials permit the physical plant to be profitably established and maintained. The main cost is for food; approximately 4,000-5,000 kg of by-catch fish are needed daily at a cost of US 20 cents per kg. If the supply of fish is inadequate, the diet is supplemented with chicken wings, legs, and necks from a slaughterhouse.

Burma

In Rangoon there are some crocodile-holding pens operated by hide buyers. It is not clear whether breeding or farming of crocodiles occurs in them or whether the operation simply acts as a clearing center for wild hides. The government has expressed interest in establishing farms in the mangrove areas near the mouth of the Irrawaddy River.

India

In 1974 an FAO report on India's crocodiles noted that the Indian gavial (*Gavialis gangeticus*) was on the verge of extinction, the saltwater crocodile was extremely rare, and the Indian mugger (*Crocodylus palustris* was a depleted, although not threatened, species.

The government, with United Nations assistance, then initiated a project for the conservation and management of all three species. This program aimed to protect and restock habitats. Animals for restocking were obtained by collecting eggs laid in the wild, incubating them under controlled conditions, raising the resulting hatchlings, and returning juveniles to specially selected sanctuaries when they reached about 1.2 m in length—at which time they are free from predation other than by man.

The project has resulted in the comeback of the gavial. By March 1979, 200 gavials had been restored to the wild. The wild population now exceeds 1,000 animals of more than 2 m length, and the number is expected to increase rapidly through natural reproduction.

The project has also carried out extensive research on crocodiles, and since its founding in 1978 the Central Crocodile Breeding and Management Training Institute, located in Hyderabad, has trained many wildlife officers in crocodile protection. Crocodile-rearing facilities are also located near Madras, Lucknow, and Cuttack. All have had success in raising the animals and restocking their habitats.

The Indian crocodile project has been a notable success and it coincided with (and perhaps helped create) a wave of local interest in India's wildlife and its conservation.

Israel

A farm stocked with American alligators (*Alligator mississippiensis*) was established at a popular hotwater spring resort area, using animals supplied by a Florida farm. The Israeli program will earn money from tourist admissions and from future production of hides. The first successful hatching of captive-bred alligators was reported in 1982.

Africa

Kenya

Near Mombasa a farm for the Nile crocodile (*Crocodylus niloticus*) has been set up to produce hides from captive-bred stock. The farm is a demonstration project of a large cement factory that is attempting to return its limestone-mined areas to productive agriculture. Some sophisticated experiments are under way on crocodile nutrition, with food for the animals produced in an intensive aquaculture project using tilapia.

Zambia

Zambia is planning a series of farms patterned after those in Zimbabwe (described below).

Zimbabwe

Zimbabwe has made great strides in captive breeding. In 1979, 87 captive females at two farms produced 1,906 eggs, and a third farm has set aside 30 captive females for breeding.

Four of the country's five crocodile farms are on the shores of Lake Kariba and the other is at Victoria Falls. The government allows each farm an annual allotment of wild eggs (averaging 2,000 to 2,500 eggs) for stocking its rearing programs. Each farm is also striving to become self-sufficient in egg production by developing successful breeding programs. The government is considering reducing each farm's allotment by the number of eggs produced annually in the farm so that each will eventually become independent of the wild populations.

Zimbabwe farmers operate on a system that obliges them to return a small percentage of live animals to the wild if the government requires it. At present, this requirement is being waived because the wild population is increasing on its own.

Zimbabwe has built its crocodile conservation program on a broad base. Crocodiles are protected throughout the country, as game animals in the country at large and as totally protected species in parks and sanctuaries. Populations have increased dramatically, from endangered status in the 1950s to over 50,000 individuals today. In the 1950s a survey of the Zambesi River and Lake Kariba revealed no crocodiles; today thousands are seen.

Zimbabwe's Department of National Parks and Wildlife is striving to ensure that its legitimate international trade in farm-raised hides does not provide illegal operators in other countries with the opportunity to sell poached hides (for example, forging papers that claim their hides originated on legitimate Zimbabwe farms). To make poaching difficult, Zimbabwe, taking a clue from the state of Louisiana, plans to use serially numbered nonremovable plastic tags to mark legitimate hides. Numbers of the tags will be noted on export permits. In addition, every export permit will be validated by the government with an engraved security stamp that is difficult to forge and that shows ink damage if any erasures or modifications are attempted. The use of such stamps is recommended by CITES, and Zimbabwe is the first nation to put them into use.

South Africa

South Africa has four crocodile farms, and another five are planned or under construction. Apart from the Natal Parks Board Crocodile Research Station at St. Lucia Estuary, which breeds Nile crocodiles for restocking and conservation purposes, all farms are for tourism and hide production. So far only one farm, outside Pretoria, is reported to produce many offspring. Only the provinces of Transvaal and Natal have wild crocodiles, and neither allows eggs, young, or adults to be collected for stocking farms. Both provinces, however, permit the killing of nuisance crocodiles on private land. Transvaal will allow one or two nuisance crocodiles to be taken captive by farmers, but it refuses permission for removing larger numbers of nuisance animals—presumably for fear that this would generate a flood of spurious nuisance complaints. Natal will not permit the removal of any wild crocodiles to farms, nor will it supply offspring from the St. Lucia station to farmers. This makes the Pretoria farm the only source of crocodiles in South Africa.

Following the example of Zimbabwe, the South African farmers (present and potential) formed a crocodile farming association in 1982.

Botswana

Several farms patterned after those in Zimbabwe are planned for the Okavango area. Petitions for approval are currently before the Botswana government.

Chad

In the late 1960s French businessmen established a farm for Nile crocodiles near Lake Chad. It collapsed after only a few years.

Ivory Coast

The government of Ivory Coast has obtained assistance from Zimbabwe to establish a conservation program for its three native crocodiles: the Nile crocodile, African slender-snouted crocodile (*Crocodylus cataphractus*), and Congo dwarf crocodile (*Osteolaemus tetraspis*). Recommendations were made for conserving the wild populations as well as for establishing farms. Field studies are under way.

Europe

Italy

A commercial farm for the South American caiman (*Caiman crocodilus*) was established in southern Italy in the late 1970s. Stock was obtained from Colombia. The animals, numbering in the thousands, arrived in Rome in winter and were transported south to the farm in an open truck. Most died from cold. Later shipments fared no better, and the few animals that survived died from poor husbandry.

The Americas

United States

There are between 15 and 20 successful alligator farms in the United States. Most are located in Florida and Louisiana, and there is at least one in California. All earn a portion of their money from tourist admissions.

Cuba

In the 1960s the Cuban government established at least two farms for crocodiles. One is located in the Zapata Peninsula National Park; the other is near Cienfuegos. The purpose of these farms is to breed crocodiles whose wetland habitat has been converted to sugarcane fields. Eventually, the farms will also produce a cash crop of hides.

Unfortunately, what started out as an admirable effort created several conservation problems because the farm managers did not realize there were two crocodiles in Cuba—the American crocodile (*Crocodylus acutus*) in brackish waters, and the Cuban crocodile (*Crocodylus rhombifer*) in freshwater areas. The two were mixed in the farms and hybridization resulted.

Mexico

The Mexican government has established several farms for Morelet's crocodile (*Crocodylus moreletii*) in Chiapas and Veracruz. The purpose is to breed the species in captivity to relieve hunting pressure on the wild population and prevent its extinction. Original funding was provided by the World Wildlife Fund. At least one of these farms still exists. Breeding has been achieved, but there have been problems of survival in hatchlings. The cause of the deaths has not been discovered.

In recent years several businessmen in Mexico have expressed interest in starting one or more crocodile farms, but none has yet materialized.

El Salvador

In the late 1960s, the Louisiana Game and Fisheries Commission supplied specimens of the American alligator to a cattle rancher in El Salvador for the purpose of establishing an experimental farm.

Louisiana was interested in studying growth rates of American alligators in a tropical nation where the animals did not have to undergo winter hibernation. The husbandry on the farm followed methods worked out in Louisiana. The animals grew fast and presumably have started breeding.

Venezuela

A captive breeding program for the Orinoco crocodile (*Crocodylus intermedius*) was established in Venezuela in the late 1970s on the ranch of Tomas Blohm. The operation is not commercial; its purpose is to prevent extinction of the species. The offspring may be used for restocking wild habitats in the future.

Under Venezuelan law there can be no commercial export of any crocodilians. In addition, the Orinoco crocodile and the various caimans are protected. However, the reptiles are everywhere killed as vermin.

Peru

Peru has proposed harvesting certain wild populations of caimans (*Caiman crocodilus*) to supply animals to a ranching operation.

Brazil

The government of Brazil is interested in establishing farms for several species of caiman, including the yacare (*Caiman crocodilus yacare*).

Other Latin American Nations

During the past two years, other Central and South American nations that have indicated their intention to set up farms for crocodiles or caimans are Bolivia, Colombia, Costa Rica, Ecuador, French Guiana, Guatemala, Guyana, Haiti, Honduras, Nicaragua, Panama, Paraguay, Suriname, and Uruguay.

Appendix **B**

Practical Crocodile Farming

This appendix is adapted from a paper by A. Pooley* that detailed the lessons learned from farming crocodiles to restock depleted habitats in Natal, South Africa. The information is presented here not as a blueprint for setting up a farm, but to show prospective farmers some of the points that they must first consider before attempting to rear crocodiles.

Farm Location

Reliable supplies of good water and suitable food are the most important considerations for establishment of a crocodile farm; the area selected must have both. Village farms also need to be close enough to wild crocodile populations for the animals to be obtained easily. Larger farms can be located farther from the source.

For small farms, a natural supply of food should also be readily available in the wild. Areas that have a fishing industry are ideal locations. For large farms, sites near slaughterhouses or fish-processing facilities are ideal.

Other considerations also include the volume of water available throughout the year, the distance over which water must be piped to the ponds, and pumping costs. The quality of the water should be established, with samples tested for salinity and acidity and, where the supply comes from mineral springs, analyzed for harmful chemicals. Chlorinated water must be tested regularly to ensure that the chlorine content is not too high, and the nature of any factory effluents present should be determined. It is important to establish whether fish, frogs, crabs, molluscs, or aquatic insects survive in the water intended for use.

Bacterial analysis is advisable where the water is drawn from a river

*Pooley, 1971.

that drains an area densely populated by humans and livestock. If the water is found to be contaminated, the stagnant pond rearing pen system should not be used, particularly when *Salmonella* spp. are present in high concentrations.

A filter system has advantages if water is pumped straight from a river carrying a heavy silt load. Apart from enabling farmers to see the animals in the pools, filtered water makes the pools and pipes easier to clean. Filtration can be achieved by drawing water from a deep pit close to the river so that the water collected seeps through sand or mud.

A reservoir or a series of supply tanks is useful as an additional method of filtering water. In the event that pumping equipment fails, such a reserve supply may prove vital to the health and survival of the crocodiles.

The ponds should receive as much sun as possible, particularly during the winter months. A series of winter air temperature recordings would be useful in choosing the site of rearing pens, since valley temperatures are often several degrees lower than the temperatures some 50 to 100 m uphill. Preference should be given to the warmer locations, taking into account the direction of local winds and heavy rains.

Soil types are the next consideration. If soils are sandy and porous, earth ponds are impractical and a concrete lining is required to retain water.

Drainage of the ponds must be carefully considered. Drainage is far easier if the ponds are built on a slight rise. Water from the ponds must not be allowed to stagnate nearby; the drainage system must be efficient. It is recommended that pens be spaced at least 8 m apart and that their drainpipes lead underground at least 10 m before emptying.

Pen Construction

Pens with rounded corners are the most successful. Crocodiles frequently choose to lie together in a pile. Square corners allow them to pile up against the angle, smothering those on the bottom and sometimes allowing animals to climb over the fence. With rounded corners, the pile cannot grow very high before the crocodiles slide sideways and the heap collapses.

Experiments in South Africa indicate that natural pools containing rooted vegetation are less prone to become sources of disease than are concrete pools. The surface of the concrete seems to become impregnated with liquid and debris from food and to become a breeding ground for bacteria. For hatchlings and very small juvenile crocodiles, concrete has an added disadvantage; its rough surface can abrade the belly skin when the animals slide in and out of the water, which can foster infection. In 5 or 10 years, even smooth concrete will erode sufficiently to become a problem.

Researchers elsewhere, however, report better results with concrete-lined ponds, which they find easier to clean. Concrete pools are useful for summer because they can be scrubbed clean and because the volume of water used is small. Normally they need only be emptied, cleaned, and refilled every third day, and there is no wastage through seepage.

The pools are best built as channels. This provides more bank for basking and enables the pools to accommodate more crocodiles. Because the larger males become belligerent only when they can see each other, floating logs, patches of grass, or channel corners are visual barriers that reduce interactions. The channel system also gives more water edge, and this appears to satisfy the territorial instinct.

Crocodiles are famous for basking in the sun, but they die surprisingly easily of heat prostration. At least one-third of the land area of a farm pen should be shaded with vegetation. The amount of space around each pool is calculated to allow ample basking room for each animal, and an area of shade must likewise be provided.

On land, crocodiles often seek contact with each other (thigmotaxis) and frequently lie piled on top of each other, but this should be a matter of choice rather than of overcrowding. There should be few enough animals in the enclosure to allow every crocodile to get out of the water if it chooses.

Ideally, only half the available number of pools should be occupied at a time, so that they can be used in rotation. In this system, the animals can be moved to fresh pools every two months (or as necessary), leaving the "used" pools to be drained and dried out to bake in the sun. After two months, the pools will then be clean and ready for use again.

An important requirement is that the pools be at least 60 cm in depth; otherwise, the water becomes too hot in summer. The pool floor should be sloped towards the drain outlet to facilitate cleaning and flushing away uneaten food. Also, the outlet pipe should be 10 cm in diameter, with a stopcock outside the enclosure, so that the pool can be cleaned and emptied efficiently. It is essential to place a screen in the drainpipe to prevent small crocodiles from escaping or being sucked out of the pool during cleaning. After some time, stagnant ponds may become difficult to clean because of the heavy growth of algae on their sides. Hard-bristle scrubbing brushes are needed to dislodge this growth. Small amounts of copper sulfate in the water will help control algae if used regularly.

The entire pond and surrounding apron must be smoothly plastered to facilitate cleaning. It helps to have a water source close to each pool from which a hose pipe can be led to pressure spray and clean the pool and its apron.

An important part of the design is a partly submerged, gently sloping ledge, some 45 cm in width, around the perimeter of the pool. This provides a shallow resting zone for the crocodiles and gives them easy access to the water. The crocodiles rest there when feeding, and the ledge prevents them from scraping their bellies and damaging their claws when they enter or leave the pool.

For small crocodiles it is advisable to roof over the entire pen with wire netting or criss-crossed strands of wire. This protects against predators. Further, young crocodiles can climb vertical wire netting with ease and will escape unless the enclosure is either roofed or has side walls that slope inwards. A skirting board (planking, sheet iron, tin, or plastic sheeting) placed against the wire netting can also prevent this. If wire netting is used for the sides of the pens it should have mesh no larger than 1 cm so that hatchlings will not injure themselves by trying to climb through. While these pools are being cleaned, care must be exercised to prevent crocodiles from falling into the empty pool.

Water can be passed continuously through the pools. The advantage of this is that during hot summer weather, when crocodiles are feeding at their maximum rate, small uneaten food particles, feces, and urine are carried away. Constant dilution of the pond's water also ensures a low bacteria level. However, the pool must be drained and scrubbed clean at least weekly.

Earthen pools are easy and cheap to build and are a "natural" habitat where vegetation can be planted and small live fish introduced; insects, frogs, and other creatures attracted to the dams will be an important addition to the diet and health of the crocodiles. Earthen pools are ideal in climates where low winter temperatures are likely to cause respiratory illness in the young animals. During cold weather the crocodiles burrow into the mudbanks and survive nights of heavy frost.

Because of the animal's burrowing capabilities, it is important to provide a strip of land 4 m wide between the pool's edge and the boundary fence. Otherwise, crocodiles may tunnel beyond the fence line. Fences must be buried at least 1 m deep to intercept the burrows and to prevent predators from burrowing in. Burrowing, however, can be hazardous, because the burrows can collapse and suffocate the animals.

In areas where the soil is porous or sandy, the floor of an earth dam can be sealed with concrete or plastic irrigation sheets. A layer of earth can be used to conceal this artificial floor. The disadvantages of earth pools are that, because of seepage, they require more water than concrete ones and that they require more maintenance because they cannot be efficiently cleaned. Even if the pools are provided with constantly circulating water, they eventually become fouled, particularly during hot weather.

Removing crocodiles from an earth pond can prove difficult, since most will take refuge in their burrows.

Capturing Crocodiles

Crocodiles are located at night, usually from a boat, by shining a light along the edges of rivers and lagoons. Because of a reflective tapetum, the eyes of crocodiles glow reddish or orange and are visible for a hundred meters or more. If the population has not become exposed to hunting and become wary of people, the animals will not submerge when the light strikes them. Dazzled by the beam, they tolerate a stealthy approach, and small animals can simply be grabbed by hand or scooped up in a net. They can then be transported in sacks to the rearing pens. Larger animals may be noosed or baited into cylindrical screen traps at places they frequent along the water's edge.

Managing a Crocodile Farm

It is necessary to sort the young crocodiles, often by size. From one clutch of eggs, some individuals will be aggressive and others may be shy or extremely timid; growth may vary from rapid to very slow, with a few individuals classed as runts.

Larger animals can be so dominant that smaller individuals will not even attempt to feed. If sorting is not done, the smaller, less-aggressive individuals do not get a fair share of the food; they grow slowly and get bitten and harassed by the larger animals. At feeding time, some will flee to the opposite side of the pen and stop feeding altogether. Keeping the young animals sorted into classes of the same size avoids many of these problems.

Nutrition and Feeding

Despite the crocodilian's reputation as a man-eater, small wild crocodiles live mainly on invertebrates and larger ones live mainly on fish. Papua New Guinea's farmers feed a varied diet of locally caught fish, crab, shrimp, frogs, snails, grasshoppers, beetles, and slaughterhouse waste. Whole animals minced up should be used, if necessary, because crocodiles require a diet of bone, intestine, scales, and other tissues to provide calcium and minerals. Bones in chopped fish must be minced thoroughly for hatchlings or very young crocodiles, or they should be fed very small fish supplemented by tadpoles or insects. One village in Papua New Guinea has shown remarkable success in rearing hatchlings on a diet of chopped fish and live freshwater shrimp.

Fish is an excellent food for the bulk feeding of a large captive population. Whole fish chopped into pieces, including the livers and hearts, forms a balanced diet that may be supplemented by meat, if available, to make up bulk. Small whole fish are particularly suitable; the crocodiles derive calcium from the bones and scales, plus roughage to facilitate digestion, while the flesh, liver, and heart are rich in nutrients and protein. The main difficulty usually lies in harvesting enough fish to meet the crocodiles' demands.

Any method of supplementing the diet with live creatures is recommended. For instance, a light can be left burning in each pen about 15 cm above the water for attracting insects. Various types of insect traps may also be used.

Crocodiles also can be fed on a variety of wastes such as offal or noncommercial fish. Ideally, a large-scale farm should be located near a poultry slaughterhouse. (Cattle offal is also satisfactory, but it is not nutritionally adequate as a sole ration for crocodiles.) Even crocodile offal itself can be fed back to crocodiles. However, the use of offal will necessitate dietary supplements to assure sufficient phosphorus and calcium. These minerals are generally provided by feeding bones to the crocodiles.

Crocodiles usually consume their food in the water, but they can also be fed on land. They will eat daily, but are able to remain active for weeks without food. If they are fed in the water of a farm pen, the water will become polluted unless there is considerable flow to carry away the debris. In extreme cases, the pools become septic. To ensure the health of the growing animals, constantly flowing water is far superior to standing water. (The Samutprakan Crocodile Farm in Thailand feeds some of its animals in water, but the small feeding pools are separate from the large regular breeding pools and at a lower level to prevent their overflowing into the breeding pools.)

It is important to feed pieces of food small enough to be swallowed without difficulty. Large fish should be cut into elongated rather than square pieces, since the bones can cause damage during swallowing. Similarly, whole live fish should not be so large that the dorsal fin may cause damage to the reptile's throat and gullet.

It is important to know the amount of food that each group of animals will consume at each meal. By feeding at the same time each day, it is easy to calculate how much is required. Moreover, the crocodiles become accustomed to a routine and the food is consumed while it is still fresh. In the hot summer months the animals will devour a full meal every 24 hours, but the feeding rate slackens with the onset of colder weather. It is then wise to start reducing frequency and quantities until food is required only every second or third day, depending on the climate. Generally, young crocodiles will refuse food when the air or water temperature falls below 60°F (15.6°C). Even in midsummer sudden cold spells may occur; at such times, it is usually futile to feed the animals or try to coax them to eat until the weather warms up again.

During hot weather conditions it is preferable to feed late in the afternoon or evenings, mainly to avoid placing the food on a hot cement surface. The food should be spread out around the edge of the pool under the shaded area so that the animals do not have to climb over one another to reach it.

In cemented pens the area where the food is laid out should be cleaned and scrubbed two hours after feeding time and any uneaten food removed from the water with a hand net. In earthen pools, the food should be placed at a different spot along the bank at each feeding. A useful aid to hygiene is keeping a few predaceous fish, such as barbel (*Clarias* spp.), in each pool to clean up scraps of uneaten food.

Population Density

Twenty-five crocodiles are considered the maximum manageable number per unit; staying within this limit reduces competition for food, bullying and fighting, and the number of injuries. A low stocking rate also results in a more even average growth rate. Most important is the fact that the overall health of the crocodiles is better than in a more crowded pen; disease problems are fewer and the symptoms easier to detect in a small group. If the units are spaced 8 m apart, there is also less danger of infectious disease spreading to other pens. The cleaning of pens is facilitated, and the disturbance caused by capturing crocodiles to be moved to other units is minimized. Housing 500 crocodiles in groups of 25 will require 20 separate pens, and an additional two pens should be provided to allow for intensive care of sick, injured, and weaker animals.

During the first year, when animals are graded frequently, they will often be moved from one pen to another. Recording the number of animals housed in each pen will make it possible to keep track of numbers and movements.

Breeding

Reproduction is impossible when crocodiles are kept in large groups composed of different species and sizes and in more or less unnatural enclosures.

Healthy, sexually mature pairs of crocodiles are usually not enough to start a breeding program. Genetic diversity to maintain a long-term breeding group must be considered, and certain environmental factors are vital for success. The distinct size and age classes of a free-living population must also be taken into consideration. Optimal sex ratios for breeding in enclosed compounds must be determined and adhered to. If a breeding unit is not based on regard for the animals' basic needs for space, nesting sites, and retreats, the larger specimens will disturb, injure, and often kill smaller specimens.

Diseases and Parasites

Disease symptoms may be easily overlooked if the observer is not familiar with the behavior of crocodiles under a variety of conditions. It is essential to know how they normally walk, swim, sleep, feed, and bask in relation to the time of day, the air and water temperatures, and the amount of sunlight or rain, by day, by night, and at different seasons of the year. Caretakers should notice the appearance of feces from healthy animals to be able to detect evidence of diarrhea, and to identify misaligned teeth and weakened limbs to detect nutritional deficiencies. Eggs are critically dependent on specific temperature and moisture requirements if the embryos are to develop normally.

It is often difficult to determine the cause of illness or death, and even if the ailment has been correctly diagnosed, it is not easy to capture and administer drugs to large numbers of sick animals. Some animals may be injured during the handling process. Emphasis on preventing disease, rather than curing it, is the best way of ensuring a healthy crop.

Almost inevitably, the water in the pools will harbor concentrations of bacteria such as salmonella. If strict hygiene is observed, however, the bacterial level will not be harmful.

It is recommended that whenever possible animals found newly dead should be dissected and vital organs such as the brain, heart, lung, liver, spleen, kidney, and stomach removed for veterinary research. Blood slides should also be taken and feces samples collected. The various specimens must be carefully labeled, frozen as quickly as possible, and packed on ice in a vacuum flask for immediate dispatch to the nearest veterinary research institute or pathologist. Alternatively, dying animals may be sent live for research purposes.

It is helpful for the handler to become thoroughly acquainted with the animal's internal anatomy, in order to distinguish between healthy and diseased organs. This knowledge, coupled with the symptoms noted before the animal dies, and the veterinary report, will be useful in future diagnosis and treatment.

One problem for the crocodile farmer is a roundworm (nematode) parasite that burrows into the belly skin. When the burrow collapses it produces an undulating track across the belly and throat scales that ruins the hide.* These parasites have been found in crocodiles from Latin

*King and Brazaitis, 1971.

America, Africa, Australia, Papua New Guinea, and Asia. They seem more prevalent in some areas and some farms than in others. The organism has been identified,* but no treatment or control has been discovered. It is, however, believed that damp, muddy conditions foster the nematode, and that to reduce it pens should have areas of dry land where the animals can bask.†

Killing, Skinning, and Tanning

Some farmers kill the crocodiles themselves, but many rear the animals and then sell them to a larger concern that is better equipped to deal with the skins. Killing is done most quickly and humanely by catching the crocodile with a noose and severing the spinal cord just behind the skull.

Many hides are ruined or severely damaged during skinning. Even a single hole resulting from a slip of the skinning knife may reduce a hide's value by 25 percent.

After skinning, the hides are normally coated with about 0.5 cm of coarse salt and rolled up. Within 48 hours they are unrolled and resalted. If the hide is not sufficiently salted, it may become infected with bacteria or fungi that cause the epidermis of the scales to decay or slip. Although this layer is removed during tannage, scale slip is a symptom of rot and usually causes damage to the finished hide product. If the decay is intense, the salted hides may become reddish or brown in color. This is called "red heat."

Although salt remains the universally used preservative for raw hides, the reptile leather industry has developed chemical fixatives that are used in addition to salt for preserving hides for tanning. Most of these pretannage fixatives are liquid and require soaking the hide in a vat, which may not be feasible in remote areas.

A pretanned hide is called a crust. It is green-gray (chrome tanned) or tan (vegetable tanned) and is stiff. The hide is dyed and glazed to its final finish. To increase the workability and to remove as many of the osteoderms as possible (if they are present), the underside of the hide is shave to an even thickness. The shaving is done by craftsmen. If they shaved too much, the hide will be thin and weak, especially over the suture between the scales.

*Ashford, R. S. and Muller, R. 1978. *Paratrichosoma* n. gen. n.sp. (Nematoda: Trichosomoididae) from the skin of the New Guinea crocodile. *Journal of Helminthology* 52:215-220.

†Information from A. Pooley.

Appendix **C**

Selected Readings

General information on the management and status of crocodiles can be found in the following:

IUCN/Crocodile Specialist Group Newsletter. Available from the editors, Peter Brazaitis and Myrna Watanabe, c/o New York Zoological Park, Bronx Zoo, The Bronx, New York 10460, USA.

A Field Guide of Captive Rearing and Management of Crocodiles in India is available from Mr. R. K. Rao, Director, Central Crocodile Breeding and Management Training Institute (Government of India), Hyderabad, Andhra Pradesh, India.

Blake, D. K. 1974. The rearing of crocodiles for commercial and conservation purposes in Rhodesia. The Rhodesia Science News 8(10):315-324.

Blake, D. K., and J. P. Loveridge. 1975. The role of commercial crocodile farming in crocodile conservation. Biological Conservation 8(4): 261-272.

Bolton, M. 1981. Crocodile Husbandry in Papua New Guinea. Field Document 4. FO: DP/PNG/74/029. Food and Agriculture Organization of the United Nations, Rome, Italy. 103 pp.

Bolton, M., and M. Laufa. 1982. The crocodile project in Papua New Guinea. Biological Conservation 22:169-179.

Brazaitis, P. (In press) Problems in the Identification of Commercial Crocodilian Hides and the Effect on Law Enforcement. Proceedings of the IUCN Crocodile Specialist Group meeting in Zimbabwe September-October, 1982. International Union for the Conservation of Nature and Natural Resources, Gland, Switzerland.

Brazaitis, P. 1973. The identification of living crocodiles. Zoologica 58(3-4):59-101.

Cott, H. B. 1954. Ecology and economic status of the crocodile in Uganda. Record of the Symposium on African Hydrobiology and Inland Fisheries, Committee on Technical Cooperation for Africa South of the Sahara 6:119-122.

Cott, H. B. 1961. Scientific results of an inquiry into the ecology and economic status of the Nile crocodile (*Crocodylus niloticus*) in Uganda and Northern Rhodesia. Transactions of the Zoological Society of London 29(4):211-337.

Downes, M. C. 1978. An explanation of the National Policy for the Crocodile Programme. Wildlife Division, Department of Lands, Surveys and Environment, Konedobu, Papua New Guinea.

Fuchs, K., nd. Die Krokodilhaut. Eduard Roether Verlag Darmstadt, Frankfurt, West Germany.

Graham, A., 1981. Mapping the pattern of crocodile nesting activity in Papua New Guinea. Field Document No. 3, FO: DP/PNG/74/029. Food and Agriculture Organization of the United Nations, Rome, Italy. 50 pp.

Joanen, T., and L. McNease. 1975. Notes on the Reproductive Biology and Captive Propagation of the American Alligator. Proceedings of the 29th Annual Conference of the Southeastern Association of Game and Fish Commissioners 29:407-415.

Joanen, T., and L. McNease. 1979. Culture of the American alligator. International Zoo Yearbook 19:61-66.

King, F. W., and P. Brazaitis. 1971. Species identification of commercial crocodile skins. Zoologica 56(2):15-72.

Loveridge, J. P. ed. 1982. The Zimbabwe Science News 16(9):196-219. Causeway, Zimbabwe. (Contains five articles on crocodilian research and conservation; from Zimbabwe, Australia, the United States, and India.)

Millichamp, N. J. 1980. Medical aspects of disease in reptile collections. In The Care and Breeding of Captive Reptiles, The British Herpetological Society, London, England.

Parker, F. 1981. New Crocodile Laws for Papua New Guinea. Division of Wildlife, Department of Lands and Environment, Konedobu, Papua New Guinea. 50 pp.

Pooley, A. C. 1969. Some observations on the rearing of crocodiles. Lammergeyer 10:45-59.

Pooley, A. C. 1971. Crocodile rearing and restocking. Pp. 104-130 in Crocodiles, IUCN Publications New Series Supplementary Paper 32. International Union for Conservation of Nature and Natural Resources, 1110 Gland, Switzerland.

Pooley, A. C. 1973. Conservation and Management of Crocodiles in Africa. Journal of the South African Wildlife Management Association 3(2):101-103.

Pooley, A. C. 1977. A report on Crocodile Farming. Papua New Guinea. Wildlife leaflet No. 77/27. pp. 1-14.

Pooley, A. C. 1981. Disappearing African Crocodiles. Oryx 16(1):38-40.

Pooley, A. C. (In press). The status of crocodiles in Africa. Paper Presented at 5th Meeting, IUCN Survival Service Commission's Crocodile Specialist Group. Gainesville, Florida, USA. September 1980.

Pooley, A. C., and C. Gans. 1976. The Nile crocodile. Scientific American 234(4):114-124.

Wallis, B. E. 1980. Market prospects for reptile leathers. Report number ITC/DIP/12, International Trade Commission, Geneva, Switzerland. 50 pp.

Wermuth, Von H., and K. Fuchs. 1978. Bestimmen von Krokodilen und ihrer Haute. Gustav Fischer Verlag, Stuttgart, New York.

Whitaker, R. 1982. Export prospects from commercial crocodile farms in Bangledesh. ITC/UNCTAD, Geneva, Switzerland. 47 pp.

Whitaker, R., and M. Kemp. 1981. The crocodile industry in Papua New Guinea: Commercial Aspects. Field Document No. 2. FO:DP/PNG/74/029. Food and Agricultural Organization of the United Nations, Rome, Italy. 35 pp.

Appendix D

Research Contacts

The following individuals are involved in crocodilian research. Most are biologists concerned with the conservation or natural history of the animals.

Australia

Applied Ecology Crocodile Farm, Edward River, Queensland 4870
Shelley Burgin, Total Environment Center, 18 Argyle Street, Sydney, New South Wales 2000
Melvin Bolton, Woodbury Farm, Woodbury Road, M.S. 142, Via Yeppoon, Queensland 4703
Max C. Downes, The Game Conservation Centre, 24 Queens Parade, North Fitzroy, Victoria 3068
Alistair Graham, P.O. Box 41266, Darwin, Northern Territory 5792
Gordon Grigg, Zoology A08, University of Sydney, Sydney, New South Wales 2006
John Lever, Koorana Crocodile Farm, M.S.F. 76, Rockhampton, Queensland 4702
Harry Messel, Head, School of Physics, University of Sydney, Sydney, New South Wales 2006
Fred Parker, 717 Ross River Road, Kirwan, Queensland 4814
Grahame Webb, School of Zoology, University of New South Wales, Kensington 2033

Bangladesh

Mohd. Reza Khan, Assistant Professor, Department of Zoology, University of Dacca, Dacca 2

Brazil

William Magnusson, Depto de Ecologia, Instituto Nacional de Pesquisas da Amazonia, Caixa Postal 478, 69000, Manaus, Amazonia

Burma

Ko Ko Gyi, Professor of Zoology, Rangoon Arts and Sciences University, Rangoon
Nyan Taw, Research Officer, People's Pearl and Fishery Corporation, Myakhwanyo Street, Thaketa, Rangoon

People's Republic of China

Huang Chu-Chien, Institute of Zoology, Chinese Academy of Science, Beijing
The Swatow Prefecture Crocodile Farm, Shantou, Guang Dong Province

Colombia

Federico Medem, Instituto Roberto Franco, Apartado Aereo 2261, Villavicencio (Meta)

Costa Rica

Gerado Budowski, Head, Natural Renewable Resources Programme, CATIE, Turrialba

Cuba

Luis Varona, Norte 29, Nuevo Vedado, La Habana 6

Dominican Republic

Jose Alberto Ottenwalder, Museo Nacional de Historia Natural, Plaza de la Cultura, Santo Domingo

Federal Republic of Germany

Karlheinz Fuchs, Schillerstrasse 2, 6257 Hunfelder 2

Guatemala

Jaime Tres I., Centro de Estudias Conservacionistas, Universidad de San Carlos de Guatemala, Avenida de la Reforma 0-63, Zona 10, Guatemala

India

B. C. Choudhury and Lala A. K. Singh, Central Crocodile Breeding and Management Training Institute, 19-4-319, Lake Dale, Rajendranagar Road, Hyderabad 500 264, Andhra Pradesh
Indian Board for Wildlife, c/o Department of Environment, Bikaner House, Shahjahan Road, New Delhi 110 011
John Sale, UNDP/FAO Crocodile Breeding and Management Project, 19-4-319, Lake Dale, Rajendranagar Road, Hyderabad 500 264, Andhra Pradesh
S. Shanmugunathan, Chief Wildlife Warden, Vivekananda Road, Coimbatore, Tamil Nadu
Sudhakar Kar, Saltwater Crocodile Research and Conservation Unit, Danginal 754 220, Via Rejkanika, District Cuttack, Orissa
Romulus Whitaker, Madras Snake Park Trust, Guindy Deer Park, Madras, Tamil Nadu, 600 002

Italy

G. S. Child, Wildlife and Parks Management Officer, FAO, Rome

Japan

Wataru Kimura, Proprietor, Atagawa Tropical and Alligator Garden, Atagawa Higashi Zlu Town, Shizuoka Prefecture

Kenya

R. D. Haller, Bamburi Portland Cement Co., Ltd., P.O. Box 90202, Mombasa

Malaysia

Paul R. Wycherley, President, Malayan Nature Society, P.O. Box 150, Kuala Lumpur

Mexico

Miguel Alvarez del Toro, Director, Instituto de Historia Natural, Departamento de Zoologia, Apartado Postal No. 6, Tuxtle Gutierrez, Chiapas, 29000
Marco Antonio Lascano B., INIREB, Apdo. No 281, Merida

Nepal

Tirtha Maskey, National Parks and Wildlife Conservation Office, Thapathali, P.O. Box 107, Kathmandu

New Zealand

Antoon de Vos, Box 34, Whitford, Auckland

Nicaragua

Milton G. Camacho B., Instituto Nicaraguense de Recursos Naturales y del Ambiente (IRENA), Depto. de Fauna Silvestre, Kim 12½ Carreterra Norte, Managua.

Pakistan

Ashiq Ahmad, Wildlife Management Specialist, Pakistan Forest Institute, Peshawar
Abdul Latif Rao, Conservator of Wildlife, National Council for Conservation of Wildlife, Ministry of Food and Agriculture, U.G. St. 51, Sector F 6/4, Islamabad

Papua New Guinea

Miro Laufa, Division of Wildlife, Department of Lands and Environment, P.O. Box 2585, Konedobu
Karol Kisokau, Director, Office of Environment and Conservation, Central Government Offices, Waigani
Navu Kwapena, Division of Wildlife, Department of Lands and Environment, Konedobu
Graham Goudie, Crocodile Farm, Lae
Martin Hollands, Monitoring Ecologist, National Crocodile Project, P.O. Box 2585 Konedobu

Peru

Pedro Vasques Ruesta, Departamento de Manejo Forestal, Universidad Nacional Agraria, Apdo. 456, La Molina

Philippines

Angel C. Alcala, Division Research, Extension and Development, Silliman University, Dumaguete City, 6501, Negros
School of Agriculture, Silliman University, Dumaguete City, 6501, Negros.

South Africa

H. B. Anthony, The Natal Parks, Game and Fish Preservation Board, Box 662, Pietermaritzburg, Natal
Antony Pooley, P.O. Box 42, St. Lucia Estuary, 3936 Zululand

Switzerland

Rene E. Honegger, Curator, Zurich Zoo, Zurichbergstrasse 221, 8044 Zurich

Thailand

U. Srisomboon, Ministry of Agriculture, Department of Livestock Development, Bangkok
Utai Youngprapakorn and Charoon Youngprapakorn, The Samutprakan Crocodile Farm and Zoo Co., Ltd., Taiban Road, Samutprakan

United Kingdom

Angus d'A. Bellairs, Department of Anatomy, St. Mary's Hospital Medical School, University of London, Paddington, London W2
H. Robert Bustard, Isle of Man

United States

John Behler, New York Zoological Society, 185th Street & Southern Blvd., The Bronx, New York 10460
Peter Brazaitis, New York Zoological Society, 185th Street & Southern Blvd., The Bronx, New York 10460
Gary Callis, Rt. 1, Box 360, Clayton, New Mexico
Robert H. Chabreck, Louisiana State University, School of Forestry and Wildlife Management, Baton Rouge, Louisiana 70803
Michael Davenport, Department of Herpetology, National Zoological Park, Smithsonian Institution, Washington, D.C. 20008
Claire Hagen, Representative, Reptile Products Association, 120 Cabrini Blvd., New York, New York 10033
Fred Hauptfuhrer, Director of Planning and Development, World Wildlife Fund, Inc., 910 17th Street NW, Washington, D.C. 20006
Howard Hunt, Atlanta Zoological Park, 518 Atlanta Ave. S.E., Atlanta, Georgia 30315
Ted Joanen, Rockefeller Wildlife Refuge, Route 1, Box 20-B, Grand Chenier, Louisiana 70643
F. Wayne King, Director, Florida State Museum, Gainesville, Florida 32611
Jeffrey W. Lang, Biology Department, University of North Dakota, University Station, Grand Forks, North Dakota 58202
Charles J. Lankester, UNDP, 1 United Nations Plaza, New York, New York 10017
James H. Powell, 1110 Kokomo Street, Plainview, Texas 79072
Charles A. Ross, Division of Reptiles and Amphibians, National Museum of Natural History, Smithsonian Institution, Washington, D.C. 20560
Myrna Watanabe, 141 Columbia Heights, Brooklyn, New York 11201

Uruguay

Federico Achaval, Departamento de Zoologia (Vertebrados), Universidad de la Republic, Cerrito 73, Montevideo

Venezuela

Tomas Blohm, Apartado 69, Caracas 1010-A
Andres Seijas Y., Servicio Nacional de Fauna Silvestre, MARNR, Apartado 184, Maracay

Zimbabwe

D. K. Blake, Department of National Parks and Wildlife Management, P.O. Box 8365, Causeway, Harare
John Loveridge, Zoology Department, University of Zimbabwe, P.O. Box MP 167, Mount Pleasant, Harare
Kevin van Jaarsveldt, Crocodile Farmers Association of Zimbabwe, P.O. Box 2569, Harare

Appendix E

Biographical Sketches of Panel Members

EDWARD S. AYENSU, Director of the Office of Biological Conservation, Smithsonian Institution, Washington, D.C., is currently the Secretary-General of the International Union of Biological Sciences. He received his B.A. in 1961 from Miami University in Ohio, M.Sc. from The George Washington University in 1963, and his Ph.D. in 1966 from the University of London. His research interests are in comparative anatomy and phylogeny of flowering plants, commercial timbers, histology of monocotyledons, economic botany, and tropical biology. An internationally recognized expert on tropical plants, he has published extensively in these areas and on topics relating to science, technology, and development, especially in developing countries. Dr. Ayensu was co-chairman of the Panel on Underexploited Tropical Plants of the Advisory Committee on Technology Innovation and chairs and serves as a member of many international bodies.

ARCHIE F. CARR, JR., is Graduate Research Professor in the Department of Zoology, University of Florida, Gainesville. As Technical Director of the Caribbean Conservation Corporation, he has directed a seasonal research program at the breeding ground of the green turtle at Tortuguero, Costa Rica, since 1952, with continuous grants from the National Science Foundation from 1955 to 1980, and has carried out investigations of marine turtle ecology and navigation in various parts of the world. The author of numerous papers, articles, and books, he received the Daniel Giraud Elliot Medal of the National Academy of Sciences for *Handbook of Turtles* and the John Burroughs Medal for *The Windward Road.* He is Research Associate of the American Museum of Natural History; Affiliate Curator of Natural Sciences, Florida State Museum; Chairman of the Marine Turtle Specialist Group of the Survival Service Commission, International Union for the Conservation of Nature; Honorary Consultant of the World Wildlife Fund; Fellow of the Linnean Society of London; Fellow of the

American Fisheries Society; and a member of Phi Beta Kappa and Sigma Xi.

In 1973 he was awarded a gold medal from the World Wildlife Fund for the application of scientific findings to the conservation of marine turtles. In 1975 he received the Edward W. Browning Award for achievement in biological conservation. In 1978 Dr. Carr was awarded the Gold Medal of the New York Zoological Society for contributions to natural science and conservation; in 1978 he became Officer of the Order of the Golden Ark (The Netherlands).

F. WAYNE KING is the Director of the Florida State Museum, Gainesville. He received a B.S. in 1957 and an M.S. in 1961 from the University of Florida and a Ph.D. from 1966 from the University of Miami. His research interests are in wildlife conservation and habitat preservation, impact of international trade on wildlife populations, and ecology and behavior of reptile populations. He worked at the New York Zoological Society from 1967 to 1975. As an international wildlife consultant, Dr. King has received honors from the Dominican Republic, the American Association of Zoological Parks and Aquariums, and from H.R.H. Prince Bernhard of The Netherlands. He has served on committees advising the State Department and the International Union for Conservation of Nature and Natural Resources on policies regarding the trade of crocodile skins, turtle products, and other wildlife materials.

FRANÇOIS MERGEN, Pinchot Professor of Forestry and Professor of Forest Genetics, Yale University, was Dean of the School of Forestry and Environmental Studies at Yale from 1965 to 1975. He received a B.A. from Luxembourg College and a B.Sc.F. from the University of New Brunswick in 1950, an M.F. in ecology in 1951, and a Ph.D. in forest genetics from Yale in 1954. He is especially knowledgeable about francophone Africa and was chairman of the Sahel program of the Board on Science and Technology for International Development and a member of the Advisory Committee on Technology Innovation. From 1960 to 1965 he was research collaborator at the Brookhaven National Laboratory. In 1966 he was the recipient of the Award for Outstanding Achievement in Biological Research from the Society of American Foresters and in 1975 was Distinguished Professor (Fulbright-Hays Program) in Yugoslavia. Before joining the Yale faculty, he served as project leader in forest genetics for the U.S. Forest Service in Florida. He has served as a consultant to FAO, various foreign governments, and private forestry companies, and he has traveled extensively in the tropical countries of Asia, Africa, and Latin America.

MICHAEL G. MORRIS is head of the Furzebrook Research Station of the Institute of Terrestrial Ecology (National Environment Research Council, U.K.). He received a B.A. in natural sciences (zoology) at the University of Cambridge of 1958, M.A. in 1962, and received his Ph.D. from London University in research on the integrated control of orchard pests. Dr. Morris worked at Monks Wood Experimental Station on the effects of grassland management on populations of invertebrates and developed a strong interest in community and applied ecology, particularly the conservation of insect populations. Recently he has become involved with problems of butterfly conservation and resource utilization. He is Secretary of the Joint Committee for the Conservation of British Insects, a Vice-Chairman of the Lepidoptera Specialist Group of IUCN's Survival Commission, and Chairman of the Habitat and Species Protection Committee of SEL (Societas Europaea Lepidoptero-Logica).

HUGH L. POPENOE is Professor of Soils, Agronomy, Botany, and Geography and Director of the Center for Tropical Agriculture and International Programs (Agriculture) at the University of Florida. He received his B.S. from the University of California at Davis in 1951 and his Ph.D. in soils from the University of Florida in 1960. His principal research interest has been in the area of tropical agriculture and land use. His early work on shifting cultivation is one of the major contributions to this sytem. He has traveled and worked in most of the countries in the tropical areas of Latin America, Asia, and Africa. He is past Chairman of the Board of Trustees of the Escuela Agricola Panamericana in Honduras, Visiting Lecturer on Tropical Public Health at the Harvard School of Public Health, and a Fellow of the American Association for the Advancement of Science, the American Society of Agronomy, the America Geographical Society, and the International Soils Science Society. He is Chairman of the Advisory Committee for Technology Innovation and a member of the Board on Science and Technology for International Development.

ROBERT MICHAEL PYLE, a writer and consulting lepidopterist based in Gray's River, Washington, has served since 1979 as Co-Compiler of the IUCN Invertebrate Red Data Book. In this capacity he is consultant to the Conservation Monitoring Centre in Cambridge, England. After receiving his B.S. and M.S. at the University of Washington, he took his Ph.D. through the School of Forestry and Environmental Studies at Yale University in 1976. He worked for the Government of Papua New Guinea on the conservation and utilization of insect resources and then with the Nature Conservancy as Northwest Land Steward. A former Fulbright Scholar to the United Kingdom, Dr. Pyle subsequently founded the Xerces Society for conservation of beneficial insects and their habitats. He has been chairman of IUCN's Lepidoptera Specialist Group (Species Survival Commission) since 1976. His publications include the *Audubon Society Field Guide to North American Butterflies.* A comprehensive book on insect conservation in his next project.

SHELDON R. SEVERINGHAUS received his Ph.D. from Cornell University in 1977 in natural resources management. He has worked on various wildlife research projects in Asia since 1964 and is representative for the Asia Foundation in Taiwan. He has published articles on butterfly conservation and wildlife industries in Taiwan, where he has been studying the butterfly and wildlife farming projects.

NOEL D. VIETMEYER, staff officer for this study, is Professional Associate of the Board on Science and Technology for International Development. A New Zealander with a Ph.D. in organic chemistry from the University of California, Berkeley, he now works on innovations in science that are important for developing countries.

Advisory Committee on Technology Innovation

HUGH POPENOE, Director, International Programs in Agriculture, University of Florida, Gainesville, Florida, *Chairman*

Members

WILLIAM BRADLEY, Consultant, New Hope, Pennsylvania

HAROLD DREGNE, Director, International Center for Arid and Semi-Arid Land Studies, Texas Tech University, Lubbock, Texas (member through 1981)

ELMER L. GADEN, JR., Department of Chemical Engineering, University of Virginia, Charlottesville, Virginia

CARL N. HODGES, Director, Environmental Research Laboratory, Tucson, Arizona

CYRUS MCKELL, Native Plants, Inc., Salt Lake City, Utah

FRANÇOIS MERGEN, Pinchot Professor of Forestry, School of Forestry and Environmental Studies, Yale University, New Haven, Connecticut (member through 1982)

DONALD L. PLUCKNETT, Consultative Group on International Agricultural Research, Washington, D.C.

THEODORE SUDIA, Deputy Science Advisor to the Secretary of Interior, Department of Interior, Washington, D.C.

Board on Science and Technology for International Development

GEORGE BUGLIARELLO, President, Polytechnic Institute of New York, Brooklyn, New York, *Chairman*

Members

SAMUEL P. ASPER, Educational Commission for Foreign Medical Graduates, Philadelphia, Pennsylvania

DAVID BELL, Department of Population Sciences, Harvard School of Public Health, Boston, Massachusetts

LEONARD BERRY, Professor, Graduate School of Geography, Clark University, Worcester, Massachusetts

ERNEST J. BRISKEY, Dean, School of Agriculture, Oregon State University, Corvallis, Oregon

HARRISON S. BROWN, Director, Resources Systems Institute, East-West Center, Honolulu, Hawaii

ROBERT H. BURRIS, Department of Biochemistry, University of Wisconsin, Madison, Wisconsin

CLAUDIA JEAN CARR, Associate Professor, Conservation and Resource Studies, University of California at Berkeley, Berkeley, California

NATE FIELDS, Director, Developing Markets, Control Data Corporation, Minneapolis, Minnesota

ROLAND J. FUCHS, Chairman, Department of Geography, University of Hawaii at Manoa, Honolulu, Hawaii

ELMER L. GADEN, JR., Department of Chemical Engineering, University of Virginia, Charlottesville, Virginia

JOHN HOWARD GIBBONS, Director, U.S. Congress, Office of Technology Assessment, Washington, D.C.

N. BRUCE HANNAY, Foreign Secretary, National Academy of Engineering, Washington, D.C.

WILLIAM HUGHES, Director, Engineering Energy Laboratory, Oklahoma State University, Stillwater, Oklahoma

WILLIAM A. W. KREBS, Vice President, Arthur D. Little, Inc., Acorn Park, Cambridge, Massachusetts

GEORGE I. LYTHCOTT, University of Wisconsin, School of Medicine, Madison, Wisconsin

JANICE E. PERLMAN, Executive Director, Committee for a New New York, New York City Partnership, New York, New York

HUGH POPENOE, Director, International Programs in Agriculture, University of Florida, Gainesville, Florida

FREDERICK C. ROBBINS, President, Institute of Medicine, National Academy of Sciences, Washington, D.C.

WALTER A. ROSENBLITH, Foreign Secretary, National Academy of Sciences, Washington, D.C.

Office of International Affairs
Board on Science and Technology for International Development (JH-217D)
National Research Council
2101 Constitution Avenue, Washington, D.C. 20418, USA

How to Order BOSTID Reports

Reports published by the Board on Science and Technology for International Development are sponsored in most instances by the U.S. Agency for International Development and are intended for free distribution primarily to readers in developing countries. A limited number of copies is available without charge to readers in the United States and other industrialized countries who are affiliated with governmental, educational, or research institutions, and who have professional interest in the subject areas treated by the report. Requests should be made on your organization's letterhead.

Single copies of published reports listed below are available free from BOSTID at the above address while the supplies last.

Energy

18. **Energy for Rural Development: Renewable Resources and Alternative Technologies for Developing Countries.** 1976. 305 pp. Examines energy technologies with power capabilities of 10-100 kW at village or rural level in terms of short- or intermediate-term availability. Identifies specific R&D efforts needed to make intermediate-term application feasible in areas offering realistic promise. (French language edition is available through NTIS, Accession No. PB 286-467.)

19. **Methane Generation from Human, Animal, and Agricultural Wastes.** 1977. 131 pp. Discusses means by which natural process of anaerobic fermentation can be controlled by man for his benefit and how the methane generated can be used as a fuel.

33. **Alcohol Fuels: Options for Developing Countries.** 1983. Examines the potential for the production and utilization of alcohol fuels in developing countries. Includes information on various tropical crops and their conversion to alcohols through both traditional and novel processes.

36. **Producer Gas: Another Fuel for Motor Transport.** 1983. During World War II Europe and Asia used wood, charcoal, and coal to fuel over a million gasoline and diesel vehicles. However, the technology has since been virtually forgotten. This report reviews producer gas and its modern potential.

38. **Supplement to Energy for Rural Development: Renewable Resources and Alternative Technologies for Developing Countries.** 1981. 240 pp. Updates the 1976 BOSTID publication and offers new material on direct and indirect uses of solar energy. Provides index to both volumes.

39. **Proceedings, International Workshop on Energy Survey Methodologies for Developing Countries.** 1980. 220 pp. Report of a 1980 workshop organized to examine past and ongoing energy survey efforts in developing countries. Includes reports from rural, urban, industry, and transportation working groups, excerpts from 12 background papers, and a directory of energy surveys for developing countries.

Technology Options for Developing Countries

8. **Ferrocement: Applications in Developing Countries.** 1973. 89 pp. Assesses state of the art and cites applications of particular interest to developing countries—boat building, construction, food and water storage facilities, etc.

14. **More Water for Arid Lands: Promising Technologies and Research Opportunities.** 1974. 153 pp. Outlines little-known but promising technologies to supply and conserve water in arid areas. (French language edition is available from BOSTID.)

21. **Making Aquatic Weeds Useful: Some Perspectives for Developing Countries.** 1976. 175 pp. Describes ways to exploit aquatic weeds for grazing, and by harvesting and processing for use as compost, animal feed, pulp, paper, and fuel. Also describes utilization for sewage and industrial wastewater treatment. Examines certain plants with potential for aquaculture.

28. **Microbial Processes: Promising Technologies for Developing Countries.** 1979. 198 pp. Discusses the potential importance of microbiology in developing countries in food and feed, plant nutrition, pest control, fuel and energy, waste treatment and utilization, and health.

31. **Food, Fuel, and Fertilizer for Organic Wastes.** 1981. 150 pp. Examines some of the opportunities for the productive utilization of organic wastes and residues commonly found in the poorer rural areas of the world.

34. **Priorities in Biotechnology Research for International Development: Proceedings of a Workshop.** 1982. 261 pp. Report of a 1982 workshop organized to examine opportunities for biotechnology research in developing countries. Includes general background papers and specific recommendations in six areas: 1) vaccines, 2) animal production, 3) monoclonal antibodies, 4) energy, 5) biological nitrogen fixation, and 6) plant cell and tissue culture.

Biological Resources

16. **Underexploited Tropical Plants with Promising Economic Value.** 1975. 187 pp. Describes 36 little-known tropical plants that, with research, could become important cash and food crops in the future. Includes cereals, roots and tubers, vegetables, fruits, oilseeds, forage plants, and others.

22. **Guayule: An Alternative Source of Natural Rubber.** 1977. 80 pp. Describes a little-known bush that grows wild in deserts of North America and produces a rubber virtually identical with that of the rubber tree. Recommends funding for guayule development.

25. **Tropical Legumes: Resources for the Future.** 1979. 331 pp. Describes plants of the family Leguminosae, including root crops, pulses, fruits, forages, timber and wood products, ornamentals, and others.

37. **The Winged Bean: A High Protein Crop for the Tropics.** (Second Edition). 1981. 59 pp. An update of BOSTID's 1975 report of this neglected tropical legume. Describes current knowledge of winged bean and its promise.

47. **Amaranth: Modern Prospects for an Ancient Crop.** 1983. Before the time of Cortez grain amaranths were staple foods of the Aztec and Inca. Today this extremely nutritious food has a bright future. The report also discusses vegetable amaranths.

Innovations in Tropical Reforestation

26. **Leucaena: Promising Forage and Tree Crop for the Tropics.** 1977. 118 pp. Describes *Leucaena leucocephala*, a little-known Mexican plant with vigorously growing, bushy types that produce nutritious forage and organic fertilizer as well as tree types that produce timber, firewood, and pulp and paper. The plant is also useful for revegetating hillslopes, providing firebreaks, and for shade and city beautification.

27. **Firewood Crops: Shrub and Tree Species for Energy Production.** 1980. 237 pp. Examines the selection of species suitable for deliberate cultivation as firewood crops in developing countries.

35. **Sowing Forests from the Air.** 1981. 64 pp. Describes experiences with establishing forests by sowing tree seed from aircraft. Suggests testing and development of the techniques for possible use where forest destructions now outpaces reforestation.

40. **Firewood Crops: Shrub and Tree Species for Energy Production.** Volume II. 1983. A continuation of BOSTID report number 27. Describes 27 species of woody plants that seem suitable candidates for fuelwood plantations in developing countries.

41. **Mangium and Other Fast-Growing Acacias for the Humid Tropics.** 1983. 63 pp. Highlights ten acacias species that are native to the tropical rain forest of Australasia. That they could become valuable forestry resources elsewhere is suggested by the exceptional performance of *Acacia mangium* in Malaysia.

42. **Calliandra: A Versatile Small Tree for the Humid Tropics.** 1983. 56 pp. This Latin American shrub is being widely planted by villagers and government agencies in Indonesia to provide firewood, prevent erosion, yield honey, and feed livestock.

43. **Casuarinas: Nitrogen-Fixing Trees for Adverse Sites.** 1983. These robust nitrogen-fixing Australasian trees could become valuable resources for planting on harsh, eroding land to provide fuel and other products. Eighteen species for tropical lowlands and highlands, temperate zones, and semiarid regions are highlighted.

Managing Tropical Animal Resources

32. **The Water Buffalo: New Prospects for an Underutilized Animal.** 1981. 118 pp. The water buffalo is performing notably well in recent trials in such unexpected places as the United States, Australia, and Brazil. Report discusses the animal's promise, particularly emphasizing its potential for use outside Asia.

44. **Butterfly Farming in Papua New Guinea.** 1983. 36 pp. Indigenous butterflies are being reared in Papua New Guinea villages in a formal government program that both provides a cash income in remote rural areas and contributes to the conservation of wildlife and tropical forests.

45. **Crocodiles as a Resource for the Tropics.** 1983. 60 pp. In most parts of the tropics crocodilian populations are being decimated, but programs in Papua New Guinea and a few other countries demonstrate that, with care, the animals can be raised for profit while the wild populations are being protected.

46. **Little-Known Asian Animals with a Promising Economic Future.** 1983. Describes banteng, madura, mithan, yak, kouprey, babirusa, Javan warty pig and other obscure, but possibly globally useful wild and domesticated animals that are indigenous to Asia.

General

29. **Postharvest Food Losses in Developing Countries.** 1978. 202 pp. Assesses potential and limitations of food-loss reduction efforts; summarizes existing work and information about losses of major food crops and fish; discusses economic and social factors involved; identifies major areas of need; and suggests policy and program options for developing countries and technical assistance agencies.

30. **U.S. Science and Technology for Development: Contributions to the UN Conference.** 1978. 226 pp. Serves the U.S. Department of State as a major background document for the U.S. national paper, 1979 United Nations Conference on Science and Technology for Development.

The following topics are now under study and will be the subjects of future BOSTID reports:

- **Leucaena: Promising Forage and Tree Crop for the Tropics** (Second Edition)
- **Jojoba**

For a complete list of publications, including those that are out of print and available only through NTIS, please write to BOSTID at the address above.

ORDER FORM

While the limited supply lasts, a free copy of *Crocodiles as a Resource for the Future* will be sent to institutionally affiliated recipients (in government, education, or research) on written request or by submission of the form below. Please indicate on the labels the names, titles, and addresses of qualified recipients and their institutions who would be interested in having this report.

Please return this form to:

Office of International Affairs (JH-217D)
National Research Council
2101 Constitution Avenue
Washington, D.C. 20418, USA

45

45

45

45

45

45